Sustainable Food Systems

In response to the challenges of a growing population and food security, there is an urgent need to construct a new agri-food sustainability paradigm. This book brings together an integrated range of key social science insights exploring the contributions and interventions necessary to build this framework. Building on over ten years of ESRC-funded theoretical and empirical research centred at BRASS, it focuses upon the key social, economic and political drivers for creating a more sustainable food system.

Themes include:

- regulation and governance
- sustainable supply chains
- public procurement
- sustainable spatial strategies associated with rural restructuring and re-calibrated urbanised food systems
- minimising biosecurity risk and animal welfare burdens.

The book critically explores the linkages between social science research and the evolving food security problems facing the world at a critical juncture in the debates associated with not only food quality, but also its provenance, vulnerability and the inherent unsustainability of current systems of production and consumption. Each chapter examines how the links between research, practice and policy can begin to contribute to more sustainable, resilient and justly distributive food systems that would be better equipped to 'feed the world' by 2050.

Terry Marsden is Director of the Sustainable Places Research Institute and Co-Director of the ESRC Centre for Business Relationships, Accountability, Sustainability & Society (BRASS), Cardiff University, UK.

Adrian Morley is Food Smart City Project Manager for Universities West Midlands, Birmingham, UK.

Other books in the Earthscan Food and Agriculture Series

Food Systems Failure
The global food crisis and the future of agriculture
Edited by Chris Rosin, Paul Stock and Hugh Campbell

Understanding the Common Agricultural Policy
By Berkeley Hill

The Sociology of Food and Agriculture
By Michael Carolan

Competition and Efficiency in International Food Supply Chains
Improving food security
By John Williams

Organic Agriculture for Sustainable Livelihoods
Edited by Niels Halberg and Adrian Muller

The Politics of Land and Food Scarcity
By Paolo De Castro, Felice Adinolfi, Fabian Capitanio, Salvatore Di Falco and Angelo Di Mambro

Principles of Sustainable Aquaculture
Promoting social, economic and environmental resilience
By Stuart Bunting

Reclaiming Food Security
By Michael S. Carolan

Food Policy in the United States
An introduction
By Parke Wilde

Precision Agriculture for Sustainability and Environmental Protection
Edited by Margaret A. Oliver, Thomas F.A. Bishop and Ben P. Marchant

Agricultural Supply Chains and the Management of Price Risk
By John Williams

Sustainable Food Systems
Building a new paradigm
Edited by Terry Marsden and Adrian Morley

For further details please visit the series page on the Routledge website:
www.routledge.com/books/series/ECEFA/

Sustainable Food Systems

Building a New Paradigm

Edited by Terry Marsden
and Adrian Morley

LONDON AND NEW YORK

from Routledge

First published 2014
by Routledge
2 Park Square, Milton Park, Abingdon, Oxon OX14 4RN

and by Routledge
711 Third Avenue, New York, NY 10017

Routledge is an imprint of the Taylor & Francis Group, an informa business

British Library Cataloguing-in-Publication Data
A catalogue record for this book is available from the British Library

Library of Congress Cataloging-in-Publication Data
Sustainable food systems : building a new paradigm / edited by Terry Marsden and Adrian Morley.
pages cm. -- (Earthscan food and agriculture)
Includes bibliographical references and index.
1. Food supply--Environmental aspects. 2. Food security. 3. Sustainable agriculture.
I. Marsden, Terry.
HD9000.5.S834 2014
338.1'9--dc23
2013028637

ISBN: 978-0-415-63954-5 (hbk)
ISBN: 978-0-203-08349-9 (ebk)

Typeset in Garamond
by GreenGate Publishing Services, Tonbridge, Kent

Printed and bound by CPI Group (UK) Ltd, Croydon, CR0 4YY

Contents

List of illustrations vii
Notes on contributors ix
Acknowledgements xiii
List of abbreviations xv

1 Current food questions and their scholarly challenges: creating and framing a sustainable food paradigm 1
TERRY MARSDEN AND ADRIAN MORLEY

2 Food futures: framing the crisis 30
ADRIAN MORLEY, JESSE MCENTEE AND TERRY MARSDEN

3 European food governance: the contrary influences of market liberalization and agricultural exceptionalism 62
ROBERT LEE

4 The public plate: harnessing the power of purchase 84
KEVIN MORGAN AND ADRIAN MORLEY

5 Sustainable food supply chains: the dynamics for change 103
ANDREW FLYNN AND KATE BAILEY

6 Biosecurity and the bioeconomy: the case of disease regulation in the UK and New Zealand 122
GARETH ENTICOTT

7 Improving animal welfare in Europe: cases of comparative bio-sustainabilities 143
MARA MIELE AND JOHN LEVER

8 **Exploring the new rural–urban interface: community food practice, land access and farmer entrepreneurialism** 166
ALEX FRANKLIN AND SELYF MORGAN

9 **The 'new frontier'? Urban strategies for food security and sustainability** 186
ROBERTA SONNINO AND JESSICA JANE SPAYDE

10 **Conclusions: building the food sustainability paradigm: research needs, complexities, opportunities** 206
TERRY MARSDEN

Index 222

Illustrations

Figures

2.1	FAO Food Price Indices, 1990–2012	32
2.2	World population estimates, 1961–2050	33
2.3	World agricultural population, 1980–2020	35
2.4	Economically active population in the agriculture, hunting, fishing and forestry industries, 1980–2005	35
2.5	Global consumption of animal products, 1961–2007	36
2.6	Share of animal-source foods in total dietary energy supplies, 1961–1963 and 2007–2009	37
2.7	Global price of maize, soybeans and wheat, 1998–2011	37
2.8	Total energy consumption by agriculture, 1980–2006	39
2.9	Crude oil prices, 1986–2012	39
2.10	World consumption of nitrogen fertilizers, 2002–2009	40
2.11	Ethanol and biodiesel production, 2000–2010	40
2.12	EU greenhouse gas emissions from agriculture, 2001–2010	41
2.13	Non-CO_2 emissions from agriculture, 2005	41
2.14	Global agricultural land per capita, 1961–2008	43
2.15	Global maize, rice and wheat yields, 1961–2012	44
2.16	Global maize, wheat, oilseed and soybean, and rice stocks, 1989–2013	44
2.17	State support for agricultural producers, 2003–2011	45
2.18	Global biotech crop production, 2011	45
2.19	Agricultural investments in low- and middle-income countries	46
2.20	Average annual foreign direct investment in agriculture	46
2.21	Area dedicated to organic agriculture, 2000–2008	47
6.1	Biosecurity objects and practices	123
6.2	Number of UK newspaper articles mentioning biosecurity	124
6.3	Bovine Tuberculosis in Great Britain and New Zealand	129

Tables

2.1 Predicted impact of climate change on agriculture 42
2.2 Recent food futures reports and their key focal points 50
2.3 Pathway to 2050: agriculture 54

Contributors

Author affiliations

Ms Kate Bailey Supply Chain Consultant, Aquity Consult Ltd.

Dr Gareth Enticott Senior Lecturer, Cardiff School of Planning and Geography, Cardiff University.

Dr Andrew Flynn Reader in Environmental Policy and Planning, Cardiff School of Planning and Geography, Cardiff University.

Dr Alex Franklin Research Fellow, Sustainable Places Research Institute, Cardiff University.

Professor Robert Lee Professor of Environmental Law, Exeter Law School, University of Exeter.

Dr John Lever Lecturer in Sustainability, University of Huddersfield Business School.

Professor Terry Marsden Professor of Environmental Policy and Planning and Director of the Sustainable Places Research Institute, Cardiff University.

Dr Jesse McEntee Managing Partner, Food Systems Research Institute, Vermont, USA.

Dr Mara Miele Reader in Human Geography, Cardiff School of Planning and Geography, Cardiff University.

Professor Kevin Morgan Professor of Governance and Development, Cardiff School of Planning and Geography, Cardiff University.

Dr Selyf Morgan Freelance Researcher, Cardiff.

Dr Adrian Morley Food Smart City Project Manager, Universities West Midlands/Harper Adams University.

Dr Roberta Sonnino Reader in Environmental Policy and Planning, Cardiff School of Planning and Geography, Cardiff University.

Ms Jessica Jane Spayde Marie Curie Research Fellow, Cardiff School of Planning and Geography, Cardiff University.

Author biographies

Kate Bailey is a Supply Chain and Business Consultant, specialising in using lean thinking to help companies improve their operations. She was previously a Senior Research Associate with Cardiff Business School and led a number of research projects across the dairy, red meat, cereals and fresh produce sectors. She also headed the research team for the Chatham House project 'UK Food Supply in the 21st Century: the New Dynamic', examining the effects of strategic future global influences on the UK food supply system. She is currently studying part-time for her PhD and her research interests include supply chain vulnerability, food supply networks and cross-chain collaboration. She was previously a Senior Research Associate with the Cardiff Business School.

Gareth Enticott is a Senior Lecturer in Rural Geography at the School of Planning and Geography, Cardiff University. His research focuses on the social impacts of animal disease, specifically in relation to bovine Tuberculosis. He has led ESRC- and Defra-funded research projects examining how farmers' understandings of animal disease shape their biosecurity practices. His most recent work examines how the governance of animal disease and veterinary expertise are shaped by organisational cultures, regulatory structures and neoliberal regimes in the UK and New Zealand.

Andrew Flynn is a Senior Lecturer in Environmental Policy and Planning at the School of Planning and Geography, Cardiff University. He has a background in policy analysis and environmental geography. His principal research interests have been in the making and delivery of policy on sustainable development, the implementation of policy and its evaluation, particularly within the food system. His work on food supply chains and technological innovations in the food system has provided insights into the relationships between the state and key economic interests and how these have led to new patterns of regulation.

Alex Franklin is a Research Fellow at the Sustainable Places Research Institute, Cardiff University, and a co-editor of the *Journal of Environmental Policy and Planning*. Her research interests include knowledge practice, community-led sustainability action, land ownership, rural development and human–animal relationships. Her recent research includes: an action research informed longitudinal study of a community food hub; a study of 'learning journeys' in connection with community sustainability action; and a collaborative study of the UK Community Land Advisory Service. Alex has also been involved with numerous international collaborative studies of community-led sustainability action, including within North and South America, mainland Europe and China.

Robert Lee is Professor of Environmental Law at the University of Exeter. He is the co-author of *The New Regulation and Governance of Food: Beyond the food* *crisis?* (Routledge, 2009) and was a researcher on the Chatham House project 'UK Food Supply in the 21st Century: the New Dynamic'. Bob also helped devise the Food Strategy for Wales: Food For Wales, Food From Wales 2010:2020. He is an Honorary Life Member of the UK Environmental Law Association (UKELA) and a Fellow of the Royal Society of Medicine.

John Lever worked from 2008 until 2013 as Researcher in the Cardiff School of Planning and Geography; he is now a Lecturer in Sustainability at the University of Huddersfield Business School. John's current research interests revolve around sustainable/resilient communities, migrant workers/ entrepreneurs and supply chain issues related to animal welfare and halal/ kosher meat markets. As a research consultant, John has been involved in work on volunteering, Roma children in education and destitution in the UK asylum system.

Terry Marsden is currently Director of Cardiff University's Sustainable Places Research Institute (PLACE) which now incorporates the UK Economic and Social Research Council supported Centre for Business Relationships, Accountability, Sustainability and Society. He also holds the established Chair of Environmental Policy and Planning in the School of Planning and Geography, Cardiff University. He has long-standing research interests in comparative agri-food studies, food security and rural development, environmental policy and sustainability science. He has published extensively in these fields and adopts an interdisciplinary approach that spans aspects of human geography, planning, rural and environmental sociology and law.

Jesse McEntee is Manager and Founder of the Food Systems Research Institute. Trained as an environmental social scientist working in the interdisciplinary settings of resource economics, land use, and regional planning, his food systems expertise draws from both private and public sectors. For instance, he supervised the intake of discovery materials for a litigation judgement of 1.6 billion dollars against Enron's energy transmission fraud in the Western US power market and has performed feasibility assessments of an international crop biofortification programme for the Bill and Melinda Gates Foundation. A prolific writer, he is the author of a number of internationally peer-reviewed publications, including articles and book chapters.

Mara Miele's research addresses the geographies of ethical food consumption and the role of animal welfare science and technology in challenging the role of farmed animals in current agricultural practices and policies. In recent years, she has focused her research on ethical food consumption, working with a large interdisciplinary network of social and animal welfare scientists to develop innovative forms of critical public engagement with science that produce the EU animal welfare standard (Welfare Quality). In 2012, Mara received the Ashby prize for the paper Miele, M. (2011) 'The taste of happiness: free-range chicken', *Environment and Planning A*, 43 (9), 2070–2090.

Kevin Morgan is Professor of Governance and Development in the School of Planning and Geography at Cardiff University, where he is also the Dean of Engagement. His food research interests revolve around public food systems, community food systems and more recently around urban food systems. He is also very active in food policy debates in his capacity as the Chair of the Bristol Food Policy Council and as a member of the UK Food Ethics Council.

Selyf Morgan has been involved in research on agri-food and rural development in a number of roles at Cardiff University. His principal interest in this area has been on how knowledge and expertise is gained by, and shared among farmers and other land managers, focusing on conversion to organic agriculture; on farmer entrepreneurship; and on the development of multifunctional agriculture. He also has an interest in processes of knowledge generation, learning and innovation in other economic sectors, and is currently working on policy evaluation; innovation and SMEs; and urban regeneration as a freelance researcher.

Adrian Morley is the Manager of the Food Smart City Project for the Universities West Midlands membership organisation and is based at Harper Adams University. Prior to this, he was a Research Associate and coordinator of food research at the Centre for Business Relationships, Accountability, Sustainability and Society, Cardiff University. His research interests focus on the transition to more sustainable food systems and in particular the role of supply chain dynamics, alternative business models and institutional demand. He has worked on a multitude of applied academic research projects in these areas and led the Cardiff team that produced the current Welsh food strategy 'Food For Wales, Food From Wales 2010:2020'.

Roberta Sonnino is a Reader in Environmental Policy and Planning in the School of Planning and Geography, Cardiff University, where she has been involved in international research on local food systems, rural development, public procurement, food security and urban food strategies. Roberta is the Director of the Research Centre for Sustainable Urban and Regional Food (SURF) and of the MSc Programme Food, Space and Society at Cardiff University.

Jessica Jane Spayde is a Marie Curie Research Fellow and PhD candidate in the PUREFOOD Programme at the School of Planning and Geography, Cardiff University. Her interest in sustainable agri-food systems brings together several research threads including sustainable public food procurement, urban and regional food systems, food policy councils, corporate social responsibility and public–private partnerships in sustainable food initiatives. Her PhD is based on a case study of the London 2012 Olympic and Paralympic Games' sustainable food initiative, and the inter-sectoral collaboration behind creating and implementing the sustainable food agenda.

Acknowledgements

We would first like to thank all of our colleagues who have contributed to this volume. The original idea and identification of a need for a book of this nature emerged out of the conference 'Researching food sustainability: reflections, challenges and the future agenda' held by the authors in February 2011, and sponsored by the ESRC Centre for Business Relationships, Accountability, Sustainability and Society (BRASS) at Cardiff University. The authors have all been active in the development of agri-food studies at Cardiff over the past decade. This period was one of considerable ferment and growth in both the breadth and depth of work in this area. It has been a process that has both engaged long-standing food researchers and also attracted new colleagues who have worked in related fields, but who have felt it important to extend their work into this inclusive and vibrant domain of scholarship.

In this sense it has been a pleasure for the editors to have worked with such a stimulating and truly interdisciplinary group of scholars. This volume is quite unique in that a large number of authors have come together in order to develop what we hope will be a valuable and lasting statement on food and its relationship with sustainability. Over the period that we have been working together, debating and discussing the food research and sustainability agenda, we have witnessed a diversification of thematic areas in which agri-food scholarship has been addressed across the world, as well as across intellectual borders. This volume is an opportunity to present some of these key themes and use them to realistically invest in the building of an inclusive, flexible and transdisciplinary sustainability paradigm.

Such an objective, of course, builds upon the work of a wide range of people and organisations, not only in the academic field, but just as importantly in the civic, public and private arenas where food issues continue to rise up policy and practice agendas. All of the authors in this publication have been heavily involved in both practice and policy as well as academic research at various times, and the role of these arenas and organisations in informing the development of academic research should be acknowledged. Funding from the UK Economic and Social Research Council (ESRC) both in BRASS as an organisation and through a range of associated research projects has been critical. Thanks should also be given to the European Commission and Welsh Government who have repeatedly sponsored food-related research at Cardiff University.

In addition, we have been pleased to host and collaborate with other significant agri-food scholars over the period of writing who have stimulated us in our work. We should particularly mention Alison Blay-Palmer, Lawrence Busch, Harriet Friedmann, Salete Cavalcante, Bill Friedland, David Goodman, Phil McMichael, John Wilkinson and Tim Lang – all frequent intellectual friends and colleagues who have contributed to many of the debates we have attempted to open up, rather than close down, in this volume.

We have also been blessed by exceptionally supportive and efficient support staff, in particular Natalie Slow, Bea Allen and Petra Anderson. Ken Peattie, as Director of BRASS, has shown continuous and generous support. Terry would also like to acknowledge Mary Anne, Joseph and Hannah, who have had to put up with many of the arguments at the dinner table, the kitchen, the garden, and on vacation. Adrian would like to thank his close friends and family, and particularly the support given by his parents Mike and Jean over recent years.

Abbreviations

AFN	agri-food network
AFS	Assured Food Standards
AHB	Animal Health Board
AHWBE	Animal Health and Welfare Board for England
BRASS	Business Relationships, Accountability, Sustainability and Society
BSE	bovine spongiform encephalopathy
bTB	bovine Tuberculosis
CAP	Common Agricultural Policy
CCT	compulsory competitive tendering
CGIAR	Consultative Group on International Agricultural Research
CSA	community supported agriculture
EFSA	European Food Safety Authority
ESRC	Economic and Social Research Council
FAO	Food and Agriculture Organization
FCFCG	Federation of City Farms and Community Gardens
FFLP	Food for Life Partnership
FMD	Foot and Mouth Disease
FPC	Food Policy Council
FPCS	food production and consumption system
GCDA	Greenwich Co-operative Development Agency
GM	Genetic Modification
IFOAM	International Federation of Organic Agriculture Movements
LCA	life cycle analysis
LFB	London Food Board
MAF	Ministry of Agriculture and Fisheries
MAFQual	MAF Quality Management
NAHAC	National Animal Health Advisory Committee
NFD	non-food diversification
NFU	National Farmers Union
NPMS	National Pest Management Strategy
OIE	Office International des Epizooties
PLACE	Sustainable Places Research Institute
RAHAC	Regional Animal Health Advice Committee

RDP	Rural Development Plan
RDR	Rural Development Regulation
REG	regional eradication group
SOE	state owned enterprise
SPS	Sanitary and Phytosanitary
SPS	Single Payment Scheme
SPTF	Sustainable Procurement Task Force
SROI	social return on investment
SURF	Sustainable Urban and Regional Food
TBAG	bTB Advisory Group
UFS	urban food strategy
VA	value-adding
WSM	Whole School Meals Ltd
WSPA	World Society for the Protection of Animals

1 Current food questions and their scholarly challenges

Creating and framing a sustainable food paradigm

Terry Marsden and Adrian Morley

Introduction: the food question in a new era

The question of food now represents one of the world's 'grand challenges'. After a prolonged period of plenty, and indeed surplus in the advanced world over the past sixty years, the food price hikes of 2008, followed by financial, fiscal and fuel crises have changed the location and significance of food policy concerns both in advanced economies and elsewhere. We have seemingly entered a new period of destabilisation that has prompted the realisation of the growing interdependence of pressures relating to the operation and governance of food 'systems' and shifting boundaries of responsibility between the state, private and civic sectors. There is a growing recognition, among scholars at least, of the need to re-examine the interconnections and linkages between food security, sustainability, sovereignty and justice in the provision, supply, allocation and consumption of food. At the same time the need to address the consequences of climate change, emerging limits to agricultural productivity, and combinative resource depletion (soils, water, phosphates, for example) in a context of growing demand and population growth, makes the environmental sustainability challenge of food all the more urgent.

This book attempts to begin to formulate a scholarly approach to the question of food by assembling contributions from leading scholars and researchers who have been studying these problems over the past two decades. Having debated the issues over a long-standing period they come together here to create a critical and open sustainability perspective on the long-standing conundrum of how to integrate food security concerns (having enough of the right food in the right place for the right people), with those of sustainability (how to create synergies of production, supply and consumption that are socially, economically and ecologically long-lasting and resilient).

This current historical conjuncture presents an important scholarly challenge for agri-food researchers. Clearly, the vibrant interdisciplinary field has expanded and diversified over the past two decades to significantly incorporate not only the critical political economy of the increasingly concentrated and global agri-food system, but also the growing 'alternative' politics and social practices of food provisioning and global fair trade (represented by the notion of Alternative Food Networks). Now, it can be argued, is the time to transcend many of these traditional binaries associated with much of that literature and create a wider and more flexible intellectual space within which to re-link the question of food to broader societal, economic, technological and political processes.

Two urgent dimensions of creative agri-food scholarship seem to be emerging through the unfolding global food crisis. First, reactions to the crisis – by governments, corporate businesses, civic and NGO groups, scientists, producers and consumers – are transcending many of the established categories of thought and action we have traditionally adhered to. These include the binaries of nature–society, urban–rural, conventional–alternative, production–consumption, and private–public interest. Second, as we will return to in the conclusion to the volume, given the centrality of food concerns in broader socially and economically sustainable development, it is increasingly necessary to examine the question of food in parallel with its wider political–economic and social conditions. In this sense the food question is a central social science question about the nature and potentiality of broader societal possibilities and contingencies relating to governance, market development, sustainable supply chains, welfare and care, and the relationships between state actions, corporate, producer and consumer actions. Food then, is not just a commodity, nor a product of 'social nature'. To address the contemporary food question we have to recognise that we are delving into wider social and multi-scalar complexities regarding the question of how different places, regions and nations can construct more sustainable futures. The food question is then an essential and significant ingredient in the creation and building of an ensemble of progressive multiple modernities that will be necessary to sustain the planet over coming decades. The contributions to this volume attempt to create a social 'lens' around which some of the key aspects of food sustainability can be viewed.

In this introductory chapter we begin to outline this framework by focusing first upon the historical development of food security and sustainability dynamics. This creates some space to explore some of the inherent problems and the associated regulation of food security and sustainability over the last two centuries. In the second part of the chapter we attempt to map out some of the contemporary key cardinal elements for a more critical, flexible, engaging and intellectually inclusive sustainability paradigm. Here, as with the succeeding contributions, we do not seek to be all encompassing or exclusive in our approach. Rather, we wish to begin to open a wider and more urgent scholarly vector within which we can more centrally locate and expose agri-food studies to major societal questions such as: how can we re-create the means and capacities to adapt and transform society beyond its carbon-based modernisation project? And, how can food actors and institutions build the means by which to create sustainable modernities?

Food security and sustainability as distinctive ecologies

Food production and consumption embodies and is affected by essential natural and metabolic processes that have historically been difficult for industry and wider forms of market and state development to control. Indeed, a continuing theme in agrarian and agri-food studies has been this long-standing 'awkwardness' of agri-food development (see Mann and Dickenson, 1978; Morgan *et al.*, 2006; Kitchen and Marsden, 2011). Unlike many other forms of economic and industrial development, food systems are distinctively embedded in physical, human, animal and plant ecologies, such that they abide by and become affected by a range of different temporal and spatial conditions, not least the long-standing disjuncture between production cycles and labour time. In framing a new sustainability paradigm for food it is necessary to recognise the continuity of this distinction, as well as the historical and long-term tendency, despite these conditions, to attempt to control and manipulate these processes by industrial and specific technological means. Agri-industrialisation in much of the twentieth century, through the applications of capital and technologies, attempted to 'smooth out' these disparities. This, however, never completely succeeded. Partly for this reason, food systems hold particular spatial and regulatory configurative features – or social ecologies – in capitalist economies. And once these configurations take hold they tend to last and influence the pathways of new dynamics. The story of food systems over the past two centuries can thus be seen as part of this paradox. No matter how globalised, technologically sophisticated or inherently footloose they may become, food systems, and their consumption and production dynamics, inherently interact with and shape spaces and places. In turn, these spaces and places – these ecologies – act to reconfigure food systems.

The first part of this chapter begins by charting this historical paradox. The historical context to the development of a sustainable food systems paradigm is important because of the current depth of crisis we now face with regard to both food sustainability and food security. We argue in the second part of the chapter that there is an urgent need to develop a new, more integrated, scientific approach to the sustainability and security of food as both of these facets experience increasingly uncontrollable problems associated with climate change, resource depletion and food insecurity. Such a scientific approach can no longer be associated with the analysis of causes alone; critically, it also has to deal with finding solutions and charting pathways out of the current crisis. We are at a historical juncture where, as we describe below, past systems of regulating the twin problems of food sustainability and food security are no longer 'fit for purpose'. They are neither ambitious nor all encompassing enough to deal with both the external and internal social and ecological costs and risks now engendered in the modern food system. Moreover, we have to recognise that there are severe 'lock-in' and denial processes at work as the crisis unfolds which partially involve the appropriation of various shades of sustainability or 'weak' forms of ecological modernisation. Indeed sustainability as a chaotic concept is everywhere; and it has been co-opted and inserted in lock-in and denial processes and discourses. The scientific quest then must now be to critically reconstruct real ecological modernisation through creating the conditions for transformative change in food systems. Central to this, as we argue here, is a re-conceptualisation and integration of the problems of food sustainability and food security.

It is not surprising therefore that, in advanced economies at least, food security and sustainability, both of which are essentially dependent upon combinations of social and natural features, have been key food governance concerns for over the last two centuries. Indeed, as industrial capitalism developed, rapid urbanisation led to more intensive enclosure of agricultural land and an ever-increasing use of fertilisers to enhance production for the growing, and increasingly urban, population. In his letters to Engels, Marx, among others in the nineteenth century, clearly understood the necessity of linking food security and sustainability by showing an appreciation for the then rapidly growing field of soil science (led by Liebig), which seemed to provide the scientific means to sustain larger and more urban settlements through the intensification of agricultural production. At that time, food security and sustainability began to find something of a long-lasting 'spatial fix', or what some Marxists called the 'metabolic rift' (see Foster, 1999). Along with various public health measures, such as clean drinking water and milk pasteurisation (see Atkins, 2010), this provided a platform to sustain continued urbanisation throughout the twentieth century.

As industrial and urban-based capitalism developed it was necessary to 'solve' the twin problems of security and sustainability, first through intensification and artificial fertilisation of land, and second, by unleashing the mechanisation of production. In the late nineteenth and early twentieth centuries these forms of agri-industrialism struggled with resolving Kautsky's formulation of the agrarian question: that is, how to continue to intensify production and appropriate some farming functions in processing and agri-industry while at the same time maintain some sort of ecological or natural balance in the agricultural transformation process (Kautsky, 1988; Goodman and Watts, 1997). Unlike other forms of industrialisation, food systems would not abide by the same principles of concentration and centralisation, partly due to the reliance upon the soil and topography, along with the dispersed nature of family farms and peasantry. However much industrial, corporate and then finance capital attempted to appropriate these 'awkward' agrarian processes – as witnessed throughout the twentieth century with the arrival of the intensive food regime (see Friedmann and McMichael, 1989) – a dominant feature of capitalist penetration in the food system has been the maintenance of family-based production units and the resilience of dispersed, land-based farming systems. The way around these 'obstacles' was to create an ever concentrated agri-industrial complex around the family farming sector, on the one hand, and subject producers to arms-length control through the operation of a continuous 'cost–price' squeeze and the dynamics of the 'technological treadmill' (Cochrane, 1968) on the other. It is perhaps remarkable how these attempted processes of subsumption of agrarian nature have been so long-lasting, despite different cycles of technological and regulatory change over the past two centuries. From early industrialisation until the present day, these 'awkward' conditions have presented many problems for the state as it has attempted to assert variable forms of public interest through its custody of food systems and compromises made with agri-industrial sectors. As we shall see, these are processes that encompass both capital and the state; for it is in both of their interests to continually attempt to arrest the problems of food security and sustainability in a global context of rapid urbanisation and population growth.

Productivism, the intensive regime and the spatial partitioning of city and countryside

Following the imperial food regime, which had gained ground in the UK and a host of settler countries during the nineteenth century based on imperial 'free trade' and settler extensive agriculture (see Marsden *et al.*, 1993), the more 'intensive food regime' that dominated the twentieth century (see Friedmann and McMichael, 1989) provided not just a major Fordist solution for the growth of cities and towns but also a clear allocation of functions for the countryside and the city (see Cronin, 1991). For instance, by the 1930s, and especially after the Second World War, the UK countryside was strictly demarcated to guarantee the stimulation of food production; at the same time, rigid restrictions were placed upon unplanned 'ribbon' development around expanding cities. In 1947, at the nadir of the British post-war financial debt crisis, and with severe food and energy shortages amidst one of the worst winters of the century, this process culminated with the passing of both the Agriculture and the Town and Country Planning Acts. The former introduced direct state subsidies for intensifying national food production, while the latter defined the rigid functionality of the 'town' and 'country' as clear regulatory spatial fixes (see Marsden *et al.*, 1993; Murdoch *et al.*, 2003).

This system, which also developed to varying degrees in other advanced countries (e.g. The Netherlands), enabled a continuous compromise to be made between food and nutritional security within a productivist model of agriculture. It also favoured a particular spatial shaping of cities, towns and villages around functional hierarchies and varying types of spatial planning mechanisms – a dynamic that is often bypassed in the literature. The relative success of this spatially regulated system meant that human health concerns regarding food could be relatively marginalised into concerns about food adulteration and minimum safety and nutritional standards (see Lang *et al.*, 2009).

As the twentieth century unfolded it became clear that there was a need for a stronger and more interventionary state to support agri-food productionism. Some commentators have talked of the establishment, from the

1930s, of an 'international food order' under US hegemony, which brought about a remarkable period of security and stability in world agricultural markets by the 1950s and 1960s (Goodman and Redclift, 1989). The US was pre-eminent among a number of settler societies in creating innovation and technologies for the period. These became integrated into an export-oriented global system. By the 1940s, the US administration was expounding a model of technological innovation and market innovation in agriculture to be disseminated internationally. This was, of course, to be matched by the Soviet race for modernisation, which encompassed state ownership of land and farms. In the West, intervention, by and large, left property rights in the hands of farmers while, through state mechanisms such as the New Deal, providing massive programmes of protection, price stabilisation, farm-income support, investment incentives, and research and development.

In the UK, as Marsden *et al.* (1993) depicted, these comprehensive systems of state intervention built around a productivist ideology provided a package of security and sustainability systems. Around a political consensus of agricultural exceptionalism (see Lee, Chapter 3, this volume) it secured land rights and land uses (between town and country), provided financial and political security for its producers and their representative bodies, and through to the 1980s provided an ideological security in protecting the national and international priority for technologically induced productivism. These productivist systems became the subject of a renaissance of scholarly work in the 1980s, both in the UK and in North America. This involved theoretical work on understanding the evolution of food regimes, as well as more empirical work on tracing the complexity of commodity chains in the increasingly globalised food systems (see Buttel and Newby, 1980; Friedland *et al.*, 1981. In summary, theoretical and empirical scholarship concentrated on four areas. These concerned, first, the ways in which capitalism sought to penetrate agriculture and the reasons why it was not always successful; second, the nature of agrarian class structures and the role of rent in providing a theoretical underpinning for a comparative political economy of agrarian class structures; third the transformation and social patterns of resistance associated with family farming; and finally a concentration upon the relations between agriculture and the state (see Buttel, 1982; Bonanno *et al.*, 1994; Goodman and Watts, 1997).

As the intensive regime of productionism continued, concentration of the non-farm parts of the food chain seemed never-ending. By the 1990s Heffernan *et al.* conceived the agri-industrial system as an increasingly globalised 'hour-glass' whereby thousands of farmers feed millions of consumers through an increasingly corporately controlled system that involves webs of interconnected input suppliers, food processors and retailers (Heffernan *et al.*, 2002). For example, by the end of the twentieth century, five major companies dominated seed production, while in the US 81 per cent of beef was processed by four firms. In addition four firms control the majority of broiler and pig production, and 81 per cent of corn exports were undertaken by three firms.

While it is important not to assume that this period of productivist regulation of food systems came to an abrupt ending in the latter parts of the twentieth century, it certainly faced a new set of challenges associated with the established synergies between security and sustainability coming under intense pressure. This became increasingly depicted in the scholarly literature, from the middle of the 1980s, when attention started to shift from critiques of productionism to wider understandings of environmentalism and consumerism.

Unravelling the productivist spatial fix: the post-productivist compromise and the dawn of the environmental crisis

Since the middle of the 1980s, the stable and productivist regulatory and state–private–public sector compromise has been progressively dismantled, and with it, much of the legitimacy of the spatial fixes that previously existed between the city and the countryside. What we have been witnessing since then is the emergence of a more volatile and contingent period of variable spatial relationships, functional uncertainties and crises of legitimacy with regard to both food security and wider questions of food sustainability.

The origins of these changes can be traced to the period between the 1980s and 2008, when deepening public health concerns associated with crises such as BSE, coupled with the recognition of a host of severe environmental 'externalities' created by the intensive food regime (see OECD, 1986; Lowe *et al.*, 1990), led to a revised regime of 'post-productionism' (see Murdoch *et al.*, 2003; Marsden, 2012). To solve these problems, faith was put in the burgeoning corporate retail sector (Marsden *et al.*, 2010; Spaargaren *et al.*, 2012) as well as in a range of usually voluntary agri-environmental schemes, without however challenging the fundamental separation between rural intensive production systems (see Buttel, 2006; van der Ploeg and Marsden, 2008) and urban consumption spaces. During this period, a myriad of new private and public food quality standards and conventions (Busch, 2007) served the function of conveniently separating, and in some cases fragmenting the growing environmental problems (e.g. sustainability) from increasing public health concerns (e.g. food and nutrition security).

This was, with hindsight and in Europe in particular, a period of the post-productivist compromise (see Marsden, 2012). As such it brought together a new set of state and spatial 'fixes'. First, growing domestic environmental and public safety concerns meant that the state had to shed some of its productivist ideology. It did this by attempting to 'ring-fence' intensive systems of production and consumption, rather than dismantling them (see Marsden, 2003). A host of agri-environmental schemes were matched by increasingly comprehensive European food safety regulation so as to constrain the producer and the food processor. Second, given the new context of 'over production', more European land was able to be 'set aside' for environmental schemes. Third, with reductions in transport costs and the rise of just-in-time logistics and 'lean' supply chain management, a greater volume and diversity of food products could be shipped and flown into the European zone from developing countries. This allowed much of intensive productionism to be relocated to the South and newly industrialising countries, while post-productivism could be fostered in the more integrated Europe. Hence, the period of post-productivism created its own set of new global spatial fixes in which it disguised productionism through new regulatory and spatial mechanisms.

While the early and indeed dominant conceptualisations of post-productivism concentrated upon the agricultural policy and regulatory change that was undoubtedly occurring in Europe from the late 1980s onwards (see Marsden and Symes, 1987; Lowe *et al*., 1993; Wilson, 2007; Ward *et al*., 2008), these tend to underplay the wider parallel shifts towards reflexive consumerism and the rise in regulatory power of the corporate retailers occurring at the same time. These forces acted to further shift economic, as well as political, power away from domestic producers and post-war productivist exceptionalism and corporatism (Self and Storing, 1962). With the rise of neoliberal applications in other economic spheres, agri-food corporatism became translated into a form of 'private-interest government' (Grant and Sargent, 1987). It is important to recognise, therefore, that the arrival of a somewhat peculiar form of European post-productivism created simultaneous conditions for the rise of national and global retailer corporatism and postmodern reflexive consumer practices. This was partly made possible in Europe by the drive for the 'European project', especially the integrated Single European Market, and the concurrent expansion of the European area to include former Eastern bloc countries. This created the political and social conditions for the emergence of new forms of collective and reflexive consumption almost as a celebration over the defeat of the (increasingly food insecure) former Eastern bloc, and the attractions of an expanding array of globalised imports and subsidised domestic products. New neoliberal and retailer-led systems of globalised (and relatively cheap) food supply were thus a powerful political and post-communist tool in late twentieth-century Europe.

As such, although post-productivism increasingly represented a regulatory shift in the sets of relationships between the state, producers and consumers, it did so by adding to the roles and responsibilities of these agents and interests rather than necessarily eradicating or negating productivism in all its forms. Post-productivism thus emerged as a compromise, in regulatory terms, to attempt to 'solve' the newly recognised EU problems of food surpluses and environmental externalities unleashed by the intensive productivist model. This was done not by eradicating intensive productivism completely, but by spatially containing its externalities at home (i.e. in the expanding EU) while stimulating and reproducing its less regulated conditions in other more distant parts of the world through highly sophisticated and neoliberalised retail-led supply chain regulation. Geopolitically, this allowed Europe to emerge as a more sophisticated and advanced food region, now more distinct from North American food hegemony on the one hand, and clearly more successful at feeding its population with an ever growing variety of fresh products than the dismantling Eastern former Soviet states on the other.

From the mid 1980s to the 2000s, and despite a series of food scares and crises, the post-productivist regime took a strong hold particularly in a more environmentally conscious Europe. This is represented by a considerable literature on the subject, especially with regard to post-productivist agricultural policy change (Wilson, 2001, 2007). However, as with the continuance of the earlier intensive productivist model, it rather acted as a way of assuaging rather than completely eradicating the twin food regulatory problems of food security and sustainability. Food security was regarded as solved, or at the very least rendered as a controllable food safety issue. Sustainability was managed through technocratic environmental management schemes that acted to 'ring fence' the problem rather than solve it. These palliatives did not eradicate the intensive regime; rather, they manipulated it into a new set of spatial fixes on a global and regional scale. Post-productivism in Europe was thus to come with some cost; a cost, not least in terms of vastly extending 'food miles', but also a cost in distantiating and exporting ecological risk and damage to other parts of the globe.

The new arena: neo-productivism, food security and the sustainability crisis

During the 2000s, however, two major changes to this global context began to redefine the meaning of both sustainability and security in the food system. First, there was the recognised arrival of 'peak food' as part-and-parcel of wider resource depletion and the dynamics of climate change. Second, the rapid growth of obesity and malnutrition in both developed and developing countries shifted the main health concerns from a series of spasmodic crises associated with food safety as part of 'risk society' (Beck, 1992), to one of human and animal biosecurity and well-being (see Rayner and Lang, 2012). Indeed, the shift from a prevailing perception of a world of food surplus to one of food deficit has been quite rapid and has gained pace since the spikes in fuel, food and energy prices that took place in 2007–2008 (see Morley *et al.*, Chapter 2, this volume). To further complicate this scenario, the last five years have also witnessed financial speculative binges, a global financial crisis and the recognised depletion of global food stocks (see McMichael, 2012) as vast productive areas have been utilised to produce biofuels rather than foods (see Mol, 2007). In addition countries such as China have become engaged in 'land grabbing' activities in Africa and Latin America in an effort to solve their internal energy, water and food security crisis (see several issues of *Journal of Agrarian Change*, 2010/11 on food security and land grabbing themes).

Notwithstanding the considerable resilience of prevailing productivist and post-productivist regimes to absorb or accommodate these growing tensions, since 2007 it has been possible to observe the ending of a twenty-year uninterrupted trend in falling food prices and the emergence of significant volatility in global food markets. What marked out 2007–2008 was not only the problem of endogenous risks prevailing in the food system, but also the new exogenous and interconnected nature of energy/resource concerns. What emerged was a resource crisis covering food, energy, water, and soil set amidst a continuing political economy of denial by corporate and national state governments still locked into the carbon arbitrage economy. These new pressures cannot be so easily spatially fixed, nor can they be contained in existing 'efficiency-led' supply chain systems. As such they represent major perturbations to existing productionist and post-productionist food systems. The literature is only just emerging on how to cope with the profundity and depth of these processes, with a reliance on headline and chart-based trends and predictive scenarios (see Ambler-Edwards *et al.*, 2009; Morley *et al.*, Chapter 2, this volume). Overall, it would seem that this is by no means a short-term 'hiccup' prior to the restoration of 'business as usual', given that the available evidence suggests that food production systems are meeting real resource limits. These are linked clearly to price volatilities in the oil and gas markets, creating a 'canary in the mine' problem for world agricultural systems.

Under these new recombinant resource pressures, whereby food systems become increasingly folded into wider energy/resource speculative 'races', it can be argued that the interests and focus of governments, whether at the global, regional or national levels, will have to shift for reasons more globally profound than those in earlier phases of post-war productivist or post-productivist food regulation. If earlier phases have proved effective and palliative whereby private and public interests, however contested, reached a level of regulatory compromise in managing security and sustainability concerns, it is clear that real and potentially irreversible social, economic and ecological 'externalities' are now being exposed. As with the nineteenth century realisation of the consequences of the 'metabolic rift', we face an at least parallel evolutionary moment in the food systems and spatial system's instability and fluidity. In this context, moreover, it is less easy to compartmentalise 'food' as a separate regulatory or system of provision, given its increasing interconnectedness to other resource sectors. As perhaps before industrialisation and recent modernisation, we have to relearn these interconnections, linking food systems to broader aspects of 'third nature' thinking (see Marsden, 2012).

The anatomy of the current food security crisis has been extensively examined (see Lee and Stokes, 2009; Lawrence *et al.*, 2010; Almås and Campbell, 2012; Spaargaren *et al.*, 2012; Maye and Kirwan, 2013). Much less attention has been devoted to the emerging landscape associated with the new variable spatial, social and economic relations and 'fixes' that local and more place-based responses to this new metabolic crisis are creating. In the next section, we will outline some of these key spatial dimensions and contingencies, which call into question the new role of cities as food policy actors.

'Third natures': from sectors to spaces: reconnecting cities with the countryside

Clearly we are left, as the second decade of the twenty-first century emerges, with a food system landscape, especially in the global North, which clings to the architecture and infrastructures of the productivist and post-productivist agri-food regimes. We are dealing with their 'sunk costs', inertias and continued spatial fixes. In this sense, to employ a transitions perspective (see Geels, 2002; Spaargaren *et al.*, 2012) we are still dealing with dominant regimes or 'worlds of food' (Morgan *et al.*, 2006) that are attempting to diffuse the new and combined 'landscape pressures' now being witnessed in different regions of the world. One major question thus becomes how will these highly contested transitions play themselves out?

One of the most vibrant trends in agri-food systems research over the past decade has been to trace the proliferation of alternative agri-food networks (AFNs) (see Goodman *et al.*, 2011 for a comprehensive synthesis of the phenomenon in North America and Europe). A major question now occurs given the growing crisis of food security and sustainability. Can alternative food movements, as a variegated assemblage of what Morgan *et al.* (2006) call the new 'moral economy' of food (Sayer, 2000) begin to scale up and out in ways that absorb more systemic and more dominant characteristics? In short, are they destined to remain an amalgam of niches or can they metamorphose into a new 'third nature' regime? As Spaargaren *et al.* (2012, p. 333) argue after completing one of the most comprehensive collections considering food system transitions:

> The foodscape of the future will be less homogenous and well-structured when compared with the post-war period. In particular, the dichotomy between alternative and mainstream food sectors and dynamics seems to have lost most of its significance. The alternative sector – be it in its primarily 'local' forms of organic agriculture or in its 'global' form of fair-trade food – is rapidly becoming more 'mainstream' both in its outlook, its major relations and dynamics and also its market shares. At the same time, the dominant, mainstream sector of global processing and retail has started to confront the (niche) challenges put forward by bottom-up, alternative food innovations in non-trivial ways, resulting in a reformulation of the dominant mainstream regime in several respects ... One does not have to be a post-modernist to recognise the fact that a sustainable, global food regime will be multi-dimensional and in some respects heterogeneous in character.

For another key commentator (McMichael, 2012, p. 117) such a critical juxtaposition is changing the nature and function of the former dominant corporate regime:

> The so called corporate food regime is a vehicle of a contradictory conjuncture, embodying a basic tension between a trajectory of 'world agriculture' represented by agro-industrialisation (food from nowhere), and a place-based form of agro-ecology (food from somewhere), including cultural survival, and expressed in food sovereignty politics – a politics of modernity in a global moral economy. That is, the food sovereignty movement is a reflex neoliberal project – seeking to reverse its catastrophic social and ecological impacts, and in so doing to develop an alternative political ontology constructed around values that are the antithesis of capital accumulation (the self valorisation of capital at whatever cost).

These two insights sum up nicely the variations of debate concerning the transitional and, at the same time, oppositional regulatory context global food systems now represent. The crisis is clearly giving vibrancy for more opposition and more heterogeneity of response which traditional regulatory governments find hard to cope with. At the same time bioeconomic advances in plant and animal genome technologies are now geared to at least a weak form of ecological modernisation, as they attempt to demonstrate how plants and animals can be intensively produced under lower carbon and chemical conditions (see Kitchen and Marsden, 2011). This is at the same time giving more oppositional vibrancy to deeper eco-economic solutions built around agro-ecology and ethical and fair trade principles (see Horlings and Marsden, 2011).

What is clear among all this fluidity and contestation is that there is less regulatory or political coherence associated with this new foodscape. Indeed, given this fluidity, 'foodscape' becomes an improved analytical tool over food 'system' or 'regime'. These new foodscapes have no similar or equivalent spatial fix. They are highly spatially variable, and as such are indeed undermining the logics of the earlier food regimes described above. For instance, the productivist regime can no longer legitimate itself without accommodating at least some ecological modernising principles. Similarly post-productionist compromises and their spatial fixes are being undermined by the depth and profundity of the ecological crisis their policies are now seen to have created. This is creating great challenges for state authorities (as we have seen in the Middle East, and increasing in China, for example) to create new legitimating frameworks that can reintegrate food security with sustainability around the needs of a neo-productivist priority (see Burton and Wilson, 2012). The effects of new economic growth and the nutrition transition in newly developing countries will necessarily lead to new innovations around neo-productivism. However, this will need new spatial as well as political compromises associated with creating synergies between ecology and economy, using different business models and sustainability approaches.

A relatively new driver for these shifts now arises from growing health and welfare concerns as part of a wider moral economy. The 'new food equation' (Morgan and Sonnino, 2010) outlined above is having major repercussions on public health at the global level. Recent data show that there are currently around 925 million undernourished people – all but 2 per cent of them in developing countries (FAO, 2010). A further 1 billion people are thought to be overweight while around 475 million people suffer from obesity. In the EU, obesity and overweight together include 60 per cent of adults and 20 per cent of children (IOTF, 2010), but the phenomenon is widespread in the global South as well (Sonnino *et al.*, 2012). Indeed, many low- and middle-income countries are experiencing rising rates of obesity and overweight and face the 'double burden' of obesity and hunger. Mexico's proportion of overweight individuals has reached 70 per cent, Brazil's 50 per cent and China's nearly 30 per cent (Cecchini *et al.*, 2010, p. 1775).

National and sectoral policies and their related spatial fixes are becoming less relevant in dealing with these problems – which, to a significant extent, have been caused by global policies that have placed too much emphasis on the production of (rather than access to) food, as explained above (see also Sonnino, 2009). In this context, a growing number of cities around the world are devising their own place-based solutions to the current security and sustainability crisis, largely (although not exclusively) through urban food strategies that aim to forge new alliances between food consumers and producers, the growing health agenda, urban centres and their surrounding rural hinterlands. This is creating a new counter-paradigm of (urban and rural) place-based eco-economic strategies that are becoming a significant counter-force to the global intensive food agenda (see Horlings and Marsden, 2012; Sonnino and Spayde, Chapter 9, this volume).

Human health and well-being are central to the narratives of many of these policy documents, especially among pioneering North American cities that have long been experiencing the negative effects of the twin food security and sustainability crises on the urban environment. Toronto is a case in point. The Canadian city, where the first food policy council was established, envisions in its recent urban food strategy a 'health-focused food system' that makes safe and nutritious food available to all urban residents, thereby nourishing the environment, protecting against climate change, promoting social justice, creating local and diverse economic development and building community (Toronto Public Health Department and Food Strategy Steering Group, 2010, p. 6). In a similar fashion, the city of Los Angeles utilises the notion of 'good food' to emphasise the centrality of citizen health within a sustainability context. Indeed, the American city's food strategy defines as a 'good food' system one that 'prioritizes the health and well being of our residents; makes healthy, high quality food affordable; contributes to a thriving economy [...]; protects and strengthens our biodiversity and natural resources throughout the region' (Los Angeles Food Policy Task Force, 2010, p. 11).

As the FAO (2011, p. 6) has recently recognised, we are now witnessing the emergence of 'a new paradigm for ecosystem-based, territorial food system planning, based on a more localised approach to food', which holds the potential to create new forms of connectivity across urban and rural landscapes, bringing the concept of sustainability into new and more profound significance as an integrative policy tool that links human and environmental health.

Towards a new conceptual architecture for a sustainable food paradigm

Clearly, given our arguments here about the historical centrality of 'solving' the problems of food security and sustainability, not least through the development of new spatial and ecological fixes, it is necessary to address how a 'new' sustainable food paradigm may begin to explore this on a long-term basis. The succeeding chapters of this volume provide more detailed and selective 'building blocks' for both the critical and normative research approaches needed. Each of the contributions is based on synoptic and synthesised accounts of long-term research conducted during the 2000s by the authors. They adopt a broad and critical sustainability perspective on the themes identified: food futures, food governance, the public realm and procurement, sustainable supply chains, biosecurity and the bioeconomy, animal welfare, and the emerging mixing and translationary processes now occurring in the new 'food equations' between urban and rural spaces. As such they begin to outline, through both critical analysis and the lens of sustainability, ways of seeing the levels of complexity and vulnerability in our current food systems; how they are emerging at the intersections between market/corporate, public and civil realms; and how they might begin to transform themselves from increasingly vulnerable frameworks of ecological and human exploitation to more resilient and sustainable systems of production and consumption. In Box 1.1, for instance, we have adapted Jacobs' (1999) conception of some of the key general parameters that are important in defining more sustainable systems. These parameters are relevant to contemporary food systems, not only by defining the pathways of travel of sustainable development, but also in distinguishing the gaps that currently exist between the realities and the potential sustainable pathways. The objective of a sustainability paradigm for food is thus not to nail down a hard and fast template along which further developments and transitions must occur. Rather, as our collegiate deliberations have exposed over many years in assembling this collection, it is about *opening up the intellectual, policy, and practice space* to think critically through new sets of sustainability 'conditions' (see Drummond and Marsden,1999; Marsden, 2003). These 'conditions' need examining critically for their 'transformational status' and ability to move beyond the current and dominant conditions of unsustainability. All of the chapters here then are explorations of these contested sets of conditions from a sustainability perspective – a perspective that re-emphasises the need to integrate questions of food security with sustainability once again. The themes covered are not meant to be exhaustive. Nevertheless, they begin to identify together some of the key dimensions of the new sustainability paradigm. We shall assess their conceptual value in more detail in the conclusion to this volume.

Box 1.1: Key parameters for defining sustainable systems

- Environment–economy integration: the eco-economy: ensuring that economic development and environmental efficiency and protection are integrated in planning and implementation.
- Futurity: an explicit concern about reducing the effects of unsustainability on the young and future generations.
- Environmental biodiversity protection and restoration: (a) environmental conservation: a recognised commitment where possible to protecting environmental resources and amenities; and (b) recognising environmental limits as a commitment to living within limits created by the 'carrying capacities' of the biosphere.
- Equity maximising, entropy minimising: a commitment to meeting at least the basic needs of the poor (relatively defined) of the present generation (as well as equity between generations).
- Quality of life and sustainable well-being: recognition that human well-being is constituted by a range of factors, including place-based abilities and capacities, and not just about income growth.
- Inclusive and multi-stakeholder capabilities and commitments for developmental and business models that are more than 'business as usual'; and ones that enhance the multiple territorial capitals of different places.

In conclusion, and by way of further introducing the sustainability paradigm, we present some of the cardinal points of departure that contextualise the more detailed treatments of the foregoing chapters. Central to this is a new commitment to integrating food security and sustainability concerns as both holistic and mutually dependent social and ecological concerns. Hence, the challenge now is not only to 'solve' these problems separately, but to combine them as a new transformatory force. Hence, unlike its productivist and post-productivist antecedents, the new sustainability paradigm can no longer afford to compartmentalise or spatially separate food security and sustainability. It has to find ways of integrating and progressing both at the same time and in the same places. The gravity of this challenge has most recently been stated in the FAO's submission to the RIO+20 summit (FAO, 2012, p. 5). It states:

> Improving agricultural and food systems is essential for a world with healthier people and healthier eco-systems. Healthy and productive lives cannot be achieved unless 'all people at all times have physical, social and economic access to sufficient, safe and nutritious food which meets their dietary needs and food preferences for an active and healthy life' (FAO, 1996). Healthy eco-systems must be resilient and productive, and provide the goods and services needed to meet current societal needs and desires without jeopardising

> the options for future generations to benefit from the full range of goods and services provided by terrestrial, aquatic and marine eco-systems. There are very strong linkages between the conditions to achieve universal food security and nutrition, responsible environmental stewardship and greater fairness in food management. They intersect in agricultural and food systems at the global, national and local levels ... FAO has three main messages for the Rio+20 summit: (i) the Rio vision of sustainable development cannot be realised unless hunger and malnutrition are eradicated; (ii) the Rio vision requires that both food consumption and production systems achieve more with less; and (iii) the transition to a sustainable future requires fundamental changes in the governance of food and agriculture and an equitable distribution of the transition costs and benefits.

The conceptual implications of such a new approach need to be recognised. We can identify the following four major conceptual hurdles that need to be addressed and overcome: Reflexive governance versus corporatist governance; the reorientation of property rights and regimes; recognising growing inequalities and income distributions and their problematic relationships with sustainability; and, finally, new spatial and governance instruments.

Reflexive versus corporatist governance

It has been argued elsewhere that progressing sustainable development, especially but not exclusively as it relates to agri-food, requires aspects of reflexive governance that reaches for and variably retains second- and third-order thinking about collective problem-solving, adaptation and the creation and maintenance of common pool resources (see Feindt, 2010). In Europe and North America particularly, many regions and cities have begun to demonstrate this with the creation of novel, innovative and overarching agri-food strategies (see Welsh Government, 2010; Sonnino and Spayde, Chapter 9, this volume). These have been inclusive in nature, bringing about much deliberation and stakeholder engagement of major public, private and civic sectors. Indeed the rapid expansion of city food strategies, food councils and charters are clearly further expressions of this tendency. These processes are reflexive also in the importance and relevance they attach to applied and engaged research and development and the expanded role of Universities in acting as loci for innovation and the engagement of regional and local actors.

At the same time however, we can also recognise the renewed insertion of a much more destructive tendency of non-reflexive governance that is expressed through a re-enforced process of neoliberalisation (see Peck *et al.*, 2011) that also gives precedence to a wider range of corporatist interests and the 'free-flow' of food goods and markets at regional and global scales (see Lee, Chapter 3, this volume; Enticott, Chapter 6, this volume). For instance, such an insertion has been dominant in the UK government since 2010 under the Conservative–Liberal coalition. Arguing that its ambitions were to be the 'greenest government ever' it has quite rapidly dismantled long-standing and hard won environmental and agri-food reflexive infrastructure by downgrading the previous government's Food Strategy 2020 to a 'delivery plan'; abolishing statutory advisory bodies such as the Royal Commission on Environmental Pollution, the Sustainable Development Commission and the Council for Rural Communities; restricting the role of the Food Standards Agency; and effectively driving a division between health and food concerns by adopting a greater focus on corporate-interest involvement. This has not just happened in the food and health areas, but also in the media, defence and finance sectors. This led a 2012 Parliamentary Select Committee enquiry (House of Commons, 2012) to conclude the need for a restated overall food strategy and a much stronger government commitment to research and development funding. The government's response to this has been to deny its own powers in influencing agri-food markets and in setting an agenda for food research.

Under these conditions it is important to delineate some of the major tenets of contemporary corporate-interest food governance, not least to create a firmer ontological basis to counter some of its principles. From a UK perspective, we can identify several key features in this regard (see Box 1.2).

Box 1.2: Features of corporatist-interest food governance in the UK post-2010

- Minimal government intervention in setting the agenda for agri-food research and development.
- Stronger 'industrial' focus and focus upon export generation (especially to 'emerging economies').
- Focus upon reducing carbon emissions through market/trading mechanisms.
- Aggregated notions of sustainable intensification or 'getting more for less'.
- Reduced regulation with a continuous 'red tape review' of environmental and agri-food sectors.
- Compartmentalisation of food, health and nutrition and a focus upon labelling and consumer food 'choice'.
- Devolving of genetically modified food adoption to 'the consumer'.
- Allowing 'the market' to decide the structure and shape of the agricultural sector.
- Reduction in the 'burden' of statutory and regional planning down to a 'new localism' agenda.
- Maximisation of green credentialism built largely on voluntarism, 'responsibility deals' with food manufacturers and retailer and fragmented projects.
- An emphasis on corporate social responsibility relating to obesity, creating new lower calorie brand opportunities for corporate retailers and caterers.

These directions are clearly attempting to reformulate food security and sustainability questions around a more bio-economic rather than eco-economic basis. There is a strong emphasis, for instance, on research and development through plant and animal genetics to create products that can produce 'more for less', even if these will struggle to be commercialised because of consumer pressure. All of the major retailers have, for instance, been allowed to withdraw their commitments to stock only non-GM-fed own-label meats. Moreover, a major feature of this approach is to continue to (i) drive a wedge between health and food questions and (ii) to marginalise and fragment real sustainability being addressed as a holistic and reflexive governance concern. One example of this is the tenuous connections made by the British government with regard to food and health. Over 60 per cent of adults are obese and a third of 10–11-year-olds are overweight or obese in the UK. The government has liaised closely with the food retailers and caterers to agree a 'calorie reduction pledge' (Department of Health, March 2012). The vast majority of the corporate commitments are new and additional, including lower calorie product lines in food and fizzy drinks, and more explicit low calorie labelling. The emphasis is again on creating a new line of product innovation for lower calorie products, rather than systematic transformation of product lines, and in providing the consumer with more 'choices' (see Flynn and Bailey, Chapter 5, this volume). This is called a 'responsibility deal' by the government and is hailed as a triumph of multi-stakeholder partnership with the food industry. It could also be called 'having your cake and eating it' by creating more product innovation and profit-making opportunities out of the growing obesity crisis. Campbell's food processors in Canada have followed the same logic in designing a new low-cost, 'nutritious' soup brand explicitly targeted at low income and food bank reliant families.

Corporate-interest governance provides a functional basis for developing these corporate-controlled new food market opportunities, whereby food security and sustainability can be compartmentalised (fixed) into a revised form of category management. Corporate-interest governance is thus a new brand of neoliberalisation; one that partially co-opts some sustainability characteristics in the name of public health and safety. It is a key vehicle for opening up more markets and legitimate trading relationships for the corporate food sector, especially at a time when food inequalities and poverty are increasing.

Reorientation of property rights and regimes

The intensification of resource exploitation is increasing as a result of the food and fuel crisis. Forms of primitive accumulation through dispossession are expressed in the global land grabbing phenomena, while in the North, and especially in the West we see a new surge of private enclosure of rural land resulting from financial and income surpluses. The destabilisation created both by the food and financial crisis – together – are creating intensive and increasingly conflictual conditions around property rights and regimes (see Coleman, 2011).

Meanwhile, the struggles continue for marginal groups, practising agro-ecology, and a range of community-led sustainability initiatives (see Franklin and Morgan, Chapter 8, this volume) and alternative food hubs (see Blay-Palmer *et al.*, 2013), to create interstitial land and infrastructure rights so as to scale up and scale out their initiatives. This is no clearer than with the explosion of urban agriculture and alternative food networks in the US and Canada (Sonnino and Spayde, Chapter 9, this volume). In Ontario for instance there is a massive growth in alternative community-based food hubs and this is starting to question the sanctity of the established green belt planning policies around cities such as Toronto (see OEDB, 2012). Any further scaling up or scaling out will run up against the established property rights associated with land use planning. Wales has recently legislated to include 'low carbon living' as a permitted development right in rural areas, while in the Netherlands and UK local opposition has been inflamed against proposals to site new mega-farm developments. All this indicates the undoing of the established spatial fixes of earlier property regimes, and the fact that pressures for new land enclosures are a key feature of the progress and contestations surrounding the sustainability paradigm.

Both these problems and more intense conflicts around both urban and rural land rights are clearly related to the first theme here of reflexive governance. The tension and management between the processes of (private) monopolisation of land rights versus (public) community rights is a critical factor in governance. This is recognised at a global scale by the FAO, which has recently approved Voluntary (note this) Guidelines on the responsible governance of tenure of land, fisheries and forests in the context of national food security (FAO, 2012). The guidelines are intended to assist states, civil society and the private sector to improve governance of tenure, contribute to alleviating hunger and poverty, empower the poor and vulnerable, enhance the environment, support national and local economic development and reform public administration. As the crisis of food security and sustainability continues to be unravelled in the North as well as the South, access to natural resources and their tenure – commonly referred to as 'land rights', but now far more ecologically multidimensional by vertically including everything from the biosphere, through the lithosphere to underlying geological formations – becomes a critical sustainability issue. Vulnerability to hunger, poor nutrition, obesity and a host of environmental effects such as climate change, flooding and drought, can partly be related to asymmetrical property rights and 'weak' forms of property regime. No clearer demonstration of these effects and to some extent their solutions through governance intervention systems exists than the recent Brazilian experience of land reform, 'zero hunger' and family farming protection linked to urban school feeding programmes. A key issue here is the government commitment to creating more flexibility and diversity in more secure and more sustainable property regimes. Here the degree of local and regional discretion can be a key issue. For instance, this is clearly the case in the US and urban growing movements. Control over property, who and how the prevailing property regime is managed, become critical questions in fostering pathways to sustainability.

Widening spatial and social inequalities and income distributions

While it is commonplace to see social inclusion as part of the 'third pillar' of sustainable development, it is less understood how the goals of sustainability can be achieved when many parts of the globe are witnessing widening social and spatial inequalities. Clearly agri-food consumption and production is bound up in this dynamic, and this needs to be recognised. It is evident that reducing income and wealth disparities is a key prerequisite for achieving real gains in both food security and sustainability. However, rather perversely we have to recognise that in many parts of the North as well as the South alternative food movements have increasingly gained their vibrancy and their membership and participation from the very act of the neoliberal state withdrawing social and food welfare safety nets. The rise of food banks in North America has institutionalised food poverty. Tackling food poverty and vulnerability is a critical aspect of inequality production and reproduction, even in countries such as Finland which have had historically high levels of social transfers and safety nets for the poor. It is thus empirically and conceptually important to link broader aspects of income disparities with the food question. In the global North this is bound up in the untangling of state responsibilities for food welfare and justice, combined with the recent relative rises in household costs of food and fuel.

It follows, therefore (see Lee and Marsden, 2011), that any move towards a real food sustainability paradigm should encompass how, even with extant forms of wide income disparities, food provisioning, procuring, consuming and producing can ameliorate some of its most extreme effects (see Morgan and Morley, Chapter 4, this volume). This points, to some extent, to the partial weakness if not redundancy of concepts such as ecological footprinting. These strictly environmental and carbon-based methodologies and their results and images obscure the emergence of food desertification and insecurity in our cities and countrysides that show the poor to hold lower ecological footprints than the heavily consuming middle and upper classes (see Alkon and Agyeman, 2011). The implicit assumption is that cutting consumption is a good and positive feature in moving and adapting to a more low-carbon future. The real question however, should be: what type of consumption? This is part of the tyranny of carbonism, in that it separates a particularly narrow carbonised definition of sustainability from the food security question, and at the same time it fails to address providence and place questions associated with food 'from somewhere'. Hence there is a conceptual need to interrelate food security with food sustainability and to link this to space and place. It is not that the sustainability paradigm should fall into the 'local trap' of always emphasising the positives of localising supply networks, but rather it is to reassess the food 'equations' and relationships between production spaces and consumption spaces. Wayne Roberts, the veteran Toronto food campaigner puts it well when he argues that we need the impoverished 'urban food consumer' to become a more knowledgeable and empowered producer of their own and their families' health and well-being (Roberts, 2008). Simultaneously we might argue that we need the 'rural producer' to become the more empowered and multifunctional bearer of sustainable goods and services for rural and urban impoverished groups (Franklin and Morgan, Chapter 8, this volume).

So far, notwithstanding some criticisms of being partly in the hands of neoliberalism (see Guthman, 2008) the alternative food movements, all in their varied and multifaceted ways, are creating new spatial and social connections. Hence, alternative food movements are a critical innovative vehicle for showing us ways of creating a real sustainable food paradigm, partly by meeting the demands of food (in)security and sustainability at the same time. They do so, also, by creating new linkages in and across urban and rural spaces, overturning our established spatial theories and supply chain models.

Hence it is necessary for the development of a sustainable food paradigm to re-centre food insecurity within the broader societal and governance context of widening social and spatial inequalities and the politics of care (see Morgan and Morley, Chapter 4, this volume), while at the same time looking at ways in which, through the necessary and diverse sustainable food transformation, we could also see alternatives to these widening and wasteful disparities. Like food waste, and ecological inefficiencies, we should conceptually treat food and income inequalities as 'wasteful' and inefficient properties.

Exploring short-termism and sustainability: temporal mechanisms in spatial policies

A final conceptual challenge for the food sustainability paradigm concerns the embedded nature of short-termism in sustainable policy making. Clearly, sustainability assumes an incorporation of the temporal as well as spatial dimensions of policy making (and un-making). As Sjöblom and Godenhjelm (2009, see also Sjöblom *et al.*, 2012) argue there are two important and somewhat contradictory processes and challenges in the era of governance being witnessed. First, the tendency for policies to be increasingly time-framed and projectified, and second, the reality in many cases that the needs of policies, and especially spatial policy interventions, demand longer-term and sometimes generational time perspectives. This is a major conundrum for the new governance regimes: how can effective policies be managed on a project-by-project basis when there is growing evidence of the need to secure long-term societal needs?

With some hindsight, while traditional twentieth-century government can be criticised for their over-burdensome hierarchical structures, excessive bureaucracies and swollen administrations, post-war agri-food policies were indeed built to last – at least for their time. Traditional productivist and some post-productivist policies were implicitly based upon stability and permanence, the more the better, while recent new governance frameworks are explicitly based upon negotiated actions, just-in-time decision making, project proliferation, fragmentation, marketisation and issue-based collaborations and programmes (see Enticott, Chapter 6, with reference to biosecurity policy, and Miele and Lever, Chapter 7, on EU animal welfare, both this volume). Clearly, institutional structures have remained, but they have been subject to far more profound internal systemic change based upon performance management principles. The problem is that while food sustainability questions demand flexibility and fluent solutions, in terms of integrated governance procedures, at the same time this can lead to mounting complexity and more or less autonomous horizontal governance processes that increasingly demand coordination and steering efforts in order to thwart disaggregation and fragmentation. Project-based governance processes also pose a threat to traditional notions of democracy: on whose mandate do the new actors operate? Who is included and who excluded? Who is accountable to whom? Furthermore, if we examine some of the central cornerstones of new governance projects and project-based interventions, it is justified to question whether they really do provide longer-term added-value to learning and long-term outcomes that contribute to real forms of sustainable development. Do they tend to fragment policy and societal actions leaving society more adrift and divided than it was in the command and control era?

Clearly, there is little chance or opportunity to retract back to that era even if there was a political appetite for doing so. It is important nevertheless to question the public efficacy of the new 'project state' in all its forms (see Sjöblom, 2009). This is particularly appropriate on two other counts. First, as Sjöblom *et al.* (2012) analyse, we now have over twenty years of practical and rich empirical experience of the 'project state' (especially in Europe), as it reaches a more mature, but one could also argue, more insecure phase. For instance, many national and international spatial policy frameworks, such as the EU LEADER and regional programmes, have been evolving over that period. But they now face new challenges with the onset of a protracted economic, political and fiscal crisis. Second, it is also clear that the project state faces new external, or landscape, pressures and threats associated with the deepening environmental and ecological crisis, both in terms of climate change and resource depletion. Here many argue (e.g. Hoff, 2011), that we are reaching tipping points in resilience, whose boundaries, once passed are irreversible. Counterpoising the critical and creative tensions between the inherent short-termism of the project state in all its forms with the growing sustainability challenge needs to be a major concern for the sustainable food paradigm. This is no more clearly expressed in this volume than in relation to the potentially powerful tool of public procurement in food analysed by Morgan and Morley (Chapter 4, this volume).

Conclusions: towards new spatial and sustainability fixes?

It is of course not clear how successful, or otherwise, the proliferation of city-region food strategies across the world are going to be in really reshaping established food systems (Sonnino and Spayde, Chapter 9, this volume; Blay-Palmer, 2010). However, it is clear that new questions are being addressed and visions created about the development of more sustainable and secure city–countryside linkages, at the same time as larger groups of consumer interests are acting to re-value consumption and production links in terms of a wider but more integrated set of security and sustainability criteria. These movements, from Porte Allegre in Brazil, to Brighton and Plymouth in the UK, raise some important questions for the role of the multilevel state and the sustainability research and development base in stimulating, scaling up and scaling out such initiatives and 'niches'. The role of public procurement and its potential to link with the preservation of small farmers and set challenging standards for institutional and household food provision (e.g. in schools and hospitals) is one key area of variable innovation.

Of equal importance are the ways in which such city initiatives are contributing to the wider realisation that current food and farming systems are not fit for purpose, and that a new approach is needed from the global level to the local level. Mirroring the IAASTD report (2009), the main EU scientific committee on agriculture, the Standing Committee on Agricultural Research (Freibauer *et al.*, 2011, p. 8) has called for radically new farming systems and research to meet these challenges, suggesting 'approaches that promise building blocks towards low-input high-output systems, integrate historical knowledge and agro-ecological principles that use nature's capacity'.

What is clear is that under the new global food crisis conditions, when deficits and surpluses both create new food equations and disrupt established spatial fixes, both between the city and the countryside on the one hand, and between the advanced and developing worlds on the other, these new city-based initiatives provide a vibrant and potentially radical approach to creating new platforms for both food security and sustainability. To become more mainstream they will require far more innovative institutional and governance support, especially at local and city-regional levels. As we know from the past, sustainable transitions in the food sector do not just occur on the 'head of a pin', they are spatially created, maintained and then reinforced. Hence the relationships between sustainable place-making and food transitions deserve to be a critical area for further sustainability science research.

A hopefully consistent theme running through this chapter has been the necessity in modern agri-food systems to attempt to 'solve' the twin conundrums associated with food security and sustainability. This was a problem in early industrialisation and urbanisation as it is today. It is important conceptually to trace these continuities, given the distinctive nature of food in capitalist development. What is striking, but sometimes omitted from debate, are the ways in which the regulation of food, its partial but nevertheless successful security and sustainability over time and space, has significantly conditioned the spatial structures and relations that are embedded in and between our societies and economies. Under the productivist and post-productivist regimes we tended to quietly ignore or at least hide these natural and physical geographies. But under the more radical crisis conditions we are currently amidst we cannot afford that luxury. In short we need to spatially as well as socially plan for more heterogeneous sustainable foodscapes in ways that engage with producers and consumers in new alliances and relationships as part of sustainable place-making.

The current food crisis, while often articulated at a global and aggregated level of discourse, will only be overcome by developing far more spatially connected as well as ecologically grounded solutions to building the adaptive capacities required. This does not espouse a defensive localism or fall into the 'local trap' of seeing local as good and global as bad. Rather it needs the innovative energy to shape global–local relations in ways that re-valorise and reconnect social ecologies.

Nevertheless, while the problematic continuities depicted here clearly exist today as they did two centuries ago, after significant bouts of state- and private-led productivism and post-productivism in the twentieth century and beyond, the problems of solving the age-old conundrums of food security and sustainability appear to have become just that bit more insurmountable.

They will, we argue, require 'new deals' for agri-food in the context of a wider sustainability paradigm. They will need civil, government and private sector support to overcome the necessary complexities in making sustainable transitions in food and health planning. They will need new engaging alliances between interdisciplinary sustainability science and the public and consumers in order to create more effective and place-based adaptive capacities. Above all, they will need new forms of effective spatial and sustainable management that harnesses the innovative potentials of a new equation between our growing cities and much needed but vulnerable countrysides. We need to re-create many Liebigs and foster a more sustainable paradigm of neo-productivism that dovetails with the growing moral and health concerns of reflexive consumers.

The contributions to this volume represent long-standing research themes that are some of the key building blocks for the new sustainable food paradigm. They are not exhaustive, but they are developed out of over a decade of interdisciplinary research conducted by Cardiff University researchers, and the ESRC research centre for Business Relationships, Accountability, Sustainability and Society at Cardiff. Each chapter tackles a major food sustainability theme, critically reviews the current state of the art and evaluates the theme in relation to developing the transformations needed to progress food sustainability. Each chapter thus represents reflexive pathways towards real sustainable transformation. While 'sustainable food' is now a wide and some might argue over-used expression, we have argued here in this introductory chapter that it is an urgent and necessary process of transformation. In researching and progressing this transformation it is clearly important to critically address the different framings of food sustainability, the degree to which these are linked to food security issues, and the degree to which the creation of a varied set of 'niche' alternatives can be mainstreamed. We are convinced that this mainstreaming is beginning to occur, but the pathways it takes, where and at what velocity it occurs are still contingencies that need to be examined. What is clear is that there is an important role for interdisciplinary, critical and engaging social science in this contested process, and that research needs blending with questions of policy and practice. When BRASS was first formed over a decade ago, it was very much in the midst of the theoretical frameworks associated with ecological modernisation. A decade later we have realised that if Huber's optimistic metaphor was correct that it is possible to transform the industrial caterpillar into the ecological butterfly, then we are still waiting for this to emerge and have real wings to fly. Over the past decade, in researching and studying food we have witnessed a more serious and profound 'turn'. This is a turn away from ecological modernisation as an option to one of complete necessity. Given the boundaries being reached with regard to the carbon model of development and the distortions in food security, it is clear that we need to reframe ecological modernisation into a wider and deeper process of sustainable transformation. The contributions here begin to provide the theoretical and empirical underpinnings of this process.

References

Alkon, A. and Agyeman, J. (eds) (2011) *Cultivating Food Justice: Race, class and sustainability*, Boston, MA: MIT Press.

Almås, R. and Campbell, H. (eds) (2012) *Rethinking Agricultural Policy Regimes: Food security, climate changes and the future resilience of global agriculture*. Research in Rural Sociology and Development Series, vol. 18, Bingley, UK: Emerald.

Ambler-Edwards, S., Bailey, K., Kiff, A., Lang, T., Lee, R., Marsden, T., Simons, D. and Tibbs, H. (2009) *Food Futures: Rethinking UK strategy*, London: Chatham House.

Atkins, P. (2010) *Liquid Materialities: A history of milk, science and the law*, Farnham: Ashgate.

Beck, U. (1992) *Risk Society: Towards a new modernity*, London: Sage.

Blay-Palmer, A. (ed.) (2010) *Re-imagining Sustainable Food Systems*, Farnham: Ashgate.

Blay-Palmer, A., Landman, K., Knezevic, I. and Hayhurst, R. (2013) 'Constructing resilient, transformative communities through sustainable "food hubs"', *Local Environment*, vol. 18, no. 5, pp. 521–528.

Bonanno, A., Busch, L., Friedland, W., Gouveia, L. and Mingione, E. (1994) (eds) *From Columbus to Conagra: The globalisation of agriculture and food*, Lawrence, KS: University of Columbia Press.

Burton, R. and Wilson, G. (2012) 'The rejuvenation of productivist agriculture: The case for cooperative neo-productivism', in R. Almås and H. Campbell (eds) *Rethinking Agricultural Policy Regimes: Food security, climate changes and the future resilience of global agriculture*. Research in Rural Sociology and Development Series, vol. 18, Bingley, UK: Emerald, pp. 51–72.

Busch, L. (2007) 'Performing the economy, performing science: From neoclassical to supply chain models in the agri-food sector', *Economy and Society*, vol. 36, no. 3, pp. 437–466.

Buttel, F. (1982) 'The political economy of agriculture in advanced industrial societies', *Current Perspectives in Social Theory*, vol. 3, pp. 27–55.

Buttel, F. (2006) 'Sustaining the unsustainable: Agri-food systems and environment in the modern world', in P. Cloke, T. Marsden and P. Mooney (eds) *Handbook of Rural Studies*, London: Sage, pp. 213–230.

Buttel, F. and Goodman, D. (1989) 'Class, state and technology and international food regimes', *Sociologia Ruralis*, vol. 14, no. 2, pp. 86–93.

Buttel, F. and Newby, H. (eds) (1980) *The Rural Sociology of Advanced Societies: Critical perspectives*, Montclair, NJ: Allenheld, Osmun.

Cecchini, M., Sassi, F., Lauer, J.A., Lee, Y.Y., Guajardo-Barron, V. and Chisholm, D. (2010), 'Tackling of unhealthy diets, physical inactivity, and obesity: Health effects and cost-effectiveness', *The Lancet*, vol. 376, no. 9754, pp. 1775–1784.

Cochrane, W. (1968, 1993) *The Development of American Agriculture: An historical analysis*, Minneapolis, MN: University of Minnesota Press.

Coleman, W.D. (2011) *Property, Territory and Globalisation: Struggles over autonomy*, Vancouver, BC: University of British Columbia Press.

Cronin, J. (1991) *Nature's Metropolis: Chicago and the Great West*, New York: W.W. Norton.

Department of Health (2012) *The Public Health Responsibility Deal*, Department of Health, London: HMSO.

Drummond, I. and Marsden, T.K. (1999) *The Condition of Sustainability*, London: Routledge.

FAO (2010) *Global Hunger Declining, but Still Unacceptably High*, Rome: FAO.

FAO (2011) *Food, Agriculture and Cities: Challenges of food and nutrition security, agriculture and ecosystem management in an urbanizing world*, FAO Food for the Cities multi-disciplinary initiative position paper, Rome: FAO.

FAO (2012) *Towards the Future We Want: End hunger and make the transition to sustainable agricultural and food systems*, FAO report submission to RIO+ 20 Summit, Rome: FAO.

Feindt, P. (2010) 'Reflexive governance of global public goods: Multi-level and multi-referential governance in agricultural policy', in E. Brosseau, T. Dedeurwaerdere and B. Siebenhüner (eds) *Reflexive Governance for Public Goods*, Boston, MA: MIT Press, pp. 159–178.

Foster, J.B. (1999) 'Marx's theory of metabolic rift: Classical foundations for environmental sociology', *American Journal of Sociology*, vol. 105, no. 2, pp. 366–405.

Freibauer, A., Mathijs, E., Brunori, G., Damianova, Z., Faroult, E., Gomis, J.G., O'Brien, L. and Treyer, S. (2011) 'Sustainable food consumption and production in a resource-constrained world: Summary findings of the EU SCAR Third Foresight Exercise', *EuroChoices*, no. 10, pp. 38–43.

Friedland, W.H. (1988) *Bologna Aftermath*, Paper presented at the Rural Studies Research Centre, UCL, London.

Friedland, H., Barton, A.E. and Thomas, R.J. (1981) *Manufacturing Green Gold: Capital, labour and technology in the lettuce industry*, New York: Cambridge University Press.

Friedmann, H. and McMichael, P. (1989) 'Agriculture and the state system', *Sociologia Ruralis*, vol. 29, no. 2, pp. 93–117.

Geels, F.W. (2002) 'Technological transitions as evolutionary reconfiguration processes: A multi-level perspective and case-study', *Research Policy*, vol. 31, no. 8/9, pp. 1257–1274.

Goodman, D. and Redclift, M. (eds) (1989) *The International Farm Crisis*, London: Macmillan.

Goodman, D. and Watts, M. (eds) (1997) *Globalising Food: Agrarian questions in global restructuring*, London: Routledge.

Goodman, D., Goodman, M. and DuPuis, M. (2011) *Alternative Food Networks: Knowledge, place and politics*, London: Routledge.

Grant, W. and Sargent, J. (1987) *Business and Politics in Britain*, Basingstoke: Macmillan.

Guthman, J. (2008) 'Neoliberalism and the making of food politics in California', *Geoforum*, vol. 39, no. 3, pp. 1171–1183.

Heffernan, W., Hendrickson, M. and Gronski, R. (2002) 'Consolidation in the food and agricultural system', *Sociologia Ruralis*, vol. 42, no. 4, pp. 347–369.

Hoff, H. (2011) *Understanding the Nexus*: Background paper for the Bonn2011 Nexus Conference, Stockholm Environment Institute, Stockholm.

Horlings, I. and Marsden, T.K. (2012) 'Exploring the "New rural paradigm" in Europe: Eco-economic strategies as a counterforce to the global competitiveness agenda', *European Urban and Regional Studies*, May 30, 2012.

House of Commons (2012) Environmental Audit Committee – Eleventh Report: Sustainable Food, Committee Report, London: HMSO.

IAASTD (2009) *Agriculture at the Crossroads*, International Assessment of Agricultural Knowledge, Science and Technology for Development, Washington DC: Island Press.

IOTF (2010) *Increasing Obesity Rates in Europe, 1985–2008*. London, UK: International Obesity Task Force.

Jacobs, M. (1999) 'Sustainable development as a contested concept', in A. Dobson (ed.) *Fairness and Futurity: Essays in sustainability and social justice*, Oxford: Oxford University Press, pp. 21–45.

Johnson, D. (ed.) (2010) 'Symposium: the 2007–2008 world food crisis' [Special issue], *Journal of Agrarian Change*, vol. 10, no. 1, pp. 1–148.

Kautsky, K. (1988, 1899) *The Agrarian Question*, London: Zwan.

Kitchen, L. and Marsden, T.K. (2011) 'Constructing sustainable communities: A theoretical exploration of the bio and eco-economy paradigms', *Local Environment*, vol. 16, no. 8, pp. 753–771.

Lang, T., Barling, D. and Caraher, M. (2009) *Food Policy: Integrating health, environment and society*, Oxford: Oxford University Press.

Lawrence, G., Lyons, K. and Wallington, T. (eds) (2010) *Food Security, Nutrition and Sustainability*, London: Earthscan.

Lee, R. and Marsden, T.K. (2011) 'Food futures: system transitions towards UK food security', *Journal of Human Rights and Environment*, vol. 2, no. 2, pp. 201–217.

Lee, R. and Stokes, E. (eds) (2009) *Economic Globalisation and Ecological Localisation: Socio-legal perspectives*, Chichester: Wiley-Blackwell.

Liebig, J. von (1859) *Letters on Modern Agriculture*, London: Walton and Maberly.

Los Angeles Food Policy Task Force (2010) *The Good Food for All agenda: Creating a new regional food system for Los Angeles*, Los Angeles, CA: City of Los Angeles.

Lowe, P., Marsden, T.K. and Whatmore, S. (eds) (1990) *Agriculture, Technology and the Rural Environment: Critical perspectives on rural change*, vol. 2, London: Fulton.

Lowe, P., Murdoch, J., Marsden, T.K., Munton, R. and Flynn, A. (1993) 'Regulating the new rural spaces: The uneven development of rural land', *Journal of Rural Studies*, vol. 9, no. 3, pp. 205–222.

McMichael, P. (2012) 'Food regime crisis and revaluing the agrarian question', in R. Almås and H. Campbell (eds) *Rethinking Agricultural Policy Regimes: Food security, climate changes and the future resilience of global agriculture*, Research in Rural Sociology and Development Series, vol. 18, Bingley: Emerald, pp. 99–122.

Mann, S. and Dickenson, J. (1978) 'Obstacles to development of a capitalist agriculture', *Journal of Peasant Studies*, vol. 5, no. 4, pp. 466–481.

Marsden, T.K. (2003) *The Condition of Rural Sustainability*, Assen, Netherlands: Royal van Gorcum.

Marsden, T.K. (2012) 'Food systems under pressure: Regulatory instabilities and the challenge of sustainable development', in G. Spaargaren, P. Oosterveer and A. Loeber (eds) *Food Practices in Transition: Changing food consumption, retail and production in the age of reflexive modernity*, Routledge Series in Sustainability Transitions, New York: Routledge, pp. 291–312.

Marsden, T.K. and Symes, D. (1987) *Survival Strategies in Post-productionist Agriculture*, Brussels: European Coordination Centre for Research and Documentation in Social Sciences.

Marsden, T.K., Lee, R., Flynn, A. and Thankappan, S. (2010) *The New Regulation and Governance of Food: Beyond the food crisis?* London: Routledge.

Marsden, T.K., Lowe, P., Murdoch, J., Munton, R. and Flynn, A. (1993) *Constructing the Countryside*, London: UCL Press.

Maye, D. and Kirwan, J. (eds) (2013) 'Food security' [Special issue], *Journal of Rural Studies*, vol. 29, pp. 1–138.

Mol, A. (2007) 'Boundless biofuels? Between environmental sustainability and vulnerability', *Sociologia ruralis*, vol. 47, no. 4, pp. 297–316.

Morgan, K. and Sonnino, R. (2010) 'The urban foodscape: World cities and the new food equation', *Cambridge Journal of Regions, Economy and Society*, vol. 3, no. 2, pp. 209–225.

Morgan, K., Marsden, T.K. and Murdoch, J. (2006) *Worlds of Food: Place, power and provenance in the food chain*, Oxford: Oxford University Press.

Murdoch, J., Lowe, P., Ward, N. and Marsden, T.K. (2003) *The Differentiated Countryside*, London: Routledge.

OECD (1986) *Agriculture and the Environment Report*, Paris: OECD.

OEDB (2012) *Golden Horse Shoe Agriculture and Food Strategy*, Toronto: Ontario Economic Development Bureau.

Peck, J., Theordore, R. and Brenner, N. (2011) 'Neo-liberalism and its malcontents', *Antipode*, vol. 41, Issue Supplement S1, pp. 94–116.

Rayner, S. and Lang, T. (2012) *Ecological Public Health*, Abingdon and New York: Routledge.

Roberts, W. (2008) *The No-nonsense Guide to World Food*, Toronto: New Internationalist Publications.

Sayer, A. (2000) 'Moral economy and political economy', *Studies in Political Economy*, vol. 62, pp. 79–104.

Self, P. and Storing, P. (1962) *The State and the Farmer*, London: Allen & Unwin.

Sjöblom, S. (ed.) (2009) 'The project state' [Special issue], *Journal of Environmental Policy and Planning*, vol. 11, no. 3, pp. 165–268.

Sjöblom, S. and Godenhjelm, S. (2009) 'Project proliferation and governance – implications for environmental management', *Journal of Environmental Policy and Planning*, vol. 11, no. 3, pp. 169–185.

Sjöblom, S., Andersson, K., Marsden, T.K and Skerratt, S. (eds) (2012) *Short-termism and Sustainability: Changing time frames in spatial policy interventions*, Farnham: Ashgate.

Sonnino, R. (2009) 'Feeding the city: Towards a new research and planning agenda', *International Planning Studies*, vol. 14, no. 4, pp. 425–436.

Sonnino, R., Spayde, J. J. and Ashe, L. (2012) 'Public policies and the construction of markets: Insights from home-grown school feeding initiatives', in F. Marques, M. Conterato and S. Schneider (eds) *Construção de Mercados para a Agricultura Familiar: Desafios para o Desenvolvimento Rural*, Porto Alegre: Federal University do Rio Grande Do Sul Press.

Spaargaren, G., Oosterveer, P. and Loeber, A. (eds) (2012) *Food Practices in Transition: Changing food consumption, retail and production in the age of reflexive modernity*, Routledge Series in Sustainability Transitions, Abingdon and New York: Routledge.

Toronto Public Health Department and Food Strategy Steering Group (2010) *Cultivating Food Connections: Toward a healthy and sustainable food system for Toronto*, Toronto: Toronto City Council.

van der Ploeg, J.D. and Marsden, T.K. (eds) (2008) *Unfolding Webs: The dynamics of regional rural development*, The Netherlands: Van Gorcum.

Ward, N., Jackson, P., Russel, P. and Wilkinson, K. (2008) 'Productivism, post-productivism and European agricultural reform: the case of sugar', *Sociologia Ruralis*, vol. 48, no. 2, pp. 118–131.

Welsh Government (2010) *Food for Wales: Food from Wales, 2010:2020*, Cardiff: Welsh Government.

Wilson, G. (2001) 'From productivism to post-productivism and back again?', *Transactions of the Institute of British Geographers*, vol. 26, no. 1, pp. 77–102.

Wilson, G. (2007) *Multifunctional Agriculture: A transition theory perspective*, Wallingford: CAB International.

2 Food futures

Framing the crisis

Adrian Morley, Jesse McEntee and Terry Marsden

Introduction: visualizing the future

Predicting the future is an often practised but seldom accurate human endeavour. History is littered with forecasts that have proved in hindsight to be widely misjudged and major events that have failed to be foreseen. The recent global financial crisis is a pertinent example of this.

The US government notoriously predicted in 1970 that by 1980 the price of crude oil 'may well decline and will in any event not experience a substantial increase', when in fact it increased by tenfold (Taleb, 2008). Despite an apparent advantage in understanding change contingent on human behaviour, social science is not immune to these difficulties. In the case of food sustainability perhaps the most famous example has been the predicted 'Malthusian' catastrophe, based on the work of Thomas Malthus in the eighteenth century, which stressed that population growth would be periodically checked by resource limitations, principally with regard to food production. Despite repeated warnings, the catastrophe has yet to materialize. With hindsight we can see that socio-technological improvements, particularly the Green Revolution and the decrease in birth rates that is associated with increased affluence, have largely prevented occurrence of the necessary conditions. Neo-Malthusian arguments continue to this day, of course, particularly with regard to environmental limits and the emerging food crisis.

Although the desire to predict the future has probably never been far from human endeavour, the objective need to do so is arguably larger now than ever. Regardless of the optimism invested in the current food system, there can be little doubt that we are facing a particular set of interconnected circumstances that are combining to cast serious uncertainty on the long-term sustainability of the current food system.

The following section outlines the current prevailing understanding of the series of resource constraints and demand factors impinging on the food system. Statistical trend data is employed where available, demonstrating the apparent direction of travel in our current global food system. This information also provides a basis for understanding how the negative impacts of these factors may be mitigated and the system adapted to deliver long-term sustainability in its broadest sense.

This is followed by a review of the recent phenomena of future food strategy developments, sparked by the commodity market spikes of 2007/8, which we are witnessing at various geographical and institutional scales and locations. An appraisal of the future of food systems, both predicted and aspired to, is of course inherent in these exercises. This chapter attempts to synthesize these documents and provide both an idea of current understanding of food futures and the possible adaptation and mitigation strategies favoured by various constituents of the food system.

Global and aggregated statistics and their trends are, clearly, only partially effective as tools for understanding the food system. The fallibility of such statistics was exemplified by the publication in October 2012 of new historical estimates by the FAO which appreciably revised down global estimates of undernourishment, albeit to still significantly high levels (FAO, WFP and IFAD, 2012). Aside from uncertainty regarding the precision and accuracy of statistical accounts, statistics mask complexities, many of which are bound up in place. Perhaps more importantly, they can obfuscate spatial and scalar inequalities in the production, trade and consumption of food (see Marsden, 2012).

Regional and local perspectives must be taken into account when considering the highly interconnected economic dynamics and trade policies as well as the complex sociocultural and socio-economic factors that affect food production and consumption systems. For example, the availability of certain farming technologies permit some populations to achieve higher levels of productivity; some trade agreements foster regionalism while others promote insularity; there is a natural variance of resources and the associated productive capability of land. Also central are consumer preferences, such as those that drive the nutrition transition (e.g. Tomlinson, 2011) and the ability (or inability) of certain people to afford food. In short, and as this book highlights, a myriad of political, economic, cultural and environmental factors must be considered to effectively understand the current, as well as future, global food system. Indeed, the ideas of locational theorist Lösch appear relevant to the case of global food futures today. He observed in 1939 that 'if everything occurred at the same time, there would be no development. If everything existed in the same place, there would be no particularity. Only space makes possible the particular, which then unfolds in time' (Lösch, 1954, p. 508).

Despite these important caveats, it is important to review the vast array of trends and perspectives that have been produced regarding the food crisis, and some of the neo-Malthusianism it has stimulated. This serves partly as a reflection of the lack of consideration of Lösch's corrective, and also to help situate how the particular global 'framing' of the crisis is currently unfolding. What is clear from the succeeding synthesis is that the interconnectedness of economic, environmental and social forces affecting current food systems is, once again, (see Marsden and Morley, Chapter 1, this volume) re-igniting the fractures and fissures inherent in food sustainability and security concerns. We therefore, at the outset to the volume, examine this process of unsustainability and growing insecurity of the food system.

Section 1: Trends and the current state of the food system

The new period of food price inflation

Nowhere is the unsustainability of the food system more visible in everyday life than in the price of food. Indeed, the price spike witnessed in 2008 propelled food security and the sustainability of the food system up the policy agenda not only in political circles but also among business and civil society. Many of the strategic food futures documents outlined in the second part of this chapter are a reaction to this event and its repercussions around the globe.

Figure 2.1 plots FAO Food Price Indices, derived as a measure of the monthly change in international prices for major food commodities, from 1990 to 2012. It shows a clear rise across all the major food groups from around the turn of the century. Food prices have continued to grow steadily since the 2007–2008 spike. Indeed, the World Bank's Food Price Index, which tracks global food commodity trade, reported further peaks in February 2011 and July 2012 (World Bank, 2012). Most analysts predict a continuation of this upward price rise under the current food system.

The impact of this price rise on consumers is direct and real. Between 2009 and 2012, the FAO Consumer Price Index for Food rose from 133 to 147 in the UK and 130 to 137 in the US (FAOSTAT, 2012). The impact on poor consumers in the developing world was both proportionally greater, due to lack of purchasing power, and generally more complex, due to greater non-market socio-economic contingencies (Compton *et al.*, 2010).

There are many factors that contribute towards short-term variations in both global commodity prices and the price paid by consumers. For example, the global recession and associated financial crisis of recent years has undoubtedly had an effect on commodity prices and consumer demand, while drought and other harvest failures can have a major impact on both current and proceeding years. The WTO Uruguay Round which commenced in 1986, and its

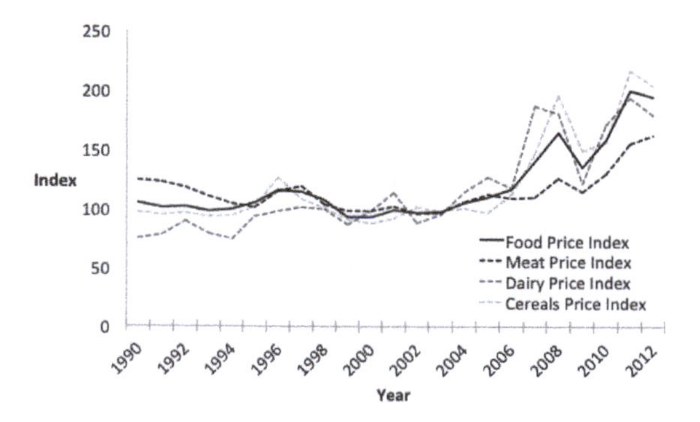

Figure 2.1 FAO Food Price Indices, 1990–2012.

Source: FAO Food Price Index.

associated set of trade reforms, impacted world food markets through reductions in surpluses and strengthening of price pressures. However, the volatility and high prices witnessed since the 1990s have been in a large part attributable to externalities unrelated to agriculture; these include developments in the energy sectors, speculative activity, and unilateral export restrictions (FAO, 2012c). There are also, however, a series of more pervasive, long-term trends, based on global resource constraints and demand shifts that are undoubtedly driving the cost of food upwards in the medium to long term.

High food prices most adversely impact the poorest populations. Estimates of global hunger remain steady at around 16 per cent and food insecurity persists in the global North as well as the South. As the volatility in food markets continues, this creates a precarious scenario for the food insecure, made worse in times of crisis when exporting countries may reduce the volume of exports through trade restrictions as a response to high prices. Food products in the lowest income countries account for nearly half of total household spending, while in the highest income countries, such as those of the OECD, this rate is around 10 per cent (IMF, 2012). Despite persistent global hunger problems, food aid has decreased significantly over the past two decades. Between 1990 and 2006, total aggregate food aid shipments of cereals decreased by 62 per cent (FAOSTAT, 2013).

Population growth and the nutrition transition

Despite periodic wars, disease and famine, the numbers of humans on the planet has grown significantly during most of the last millennia, largely due to technological and social developments combined with the exponential nature of population growth. As Figure 2.2 illustrates, in the last fifty years, the global population is thought to have grown from around 3 billion to current estimates of around 7 billion. Of these 7 billion, according to some, as many as 1 billion are undernourished, 1 billion are overweight, and 1 billion malnourished (FAO, 2009; Beddington *et al.*, 2011). Estimates for future growth vary considerably but the FAO predicts a likely figure of around 9 billion by about 2050.

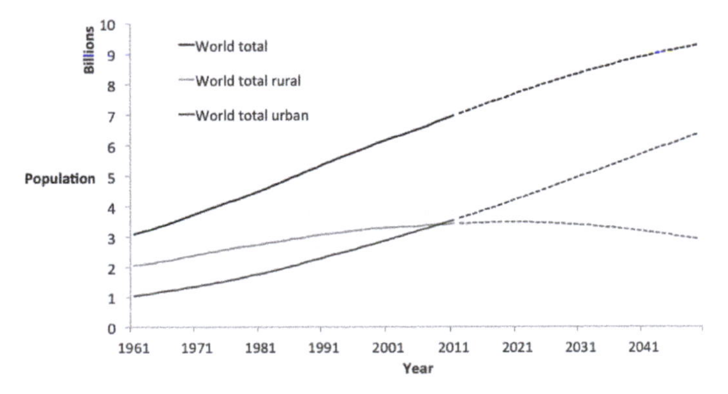

Figure 2.2 World population estimates, 1961–2050.

Source: FAOSTAT Population Data.

Within this global picture there is a clear distinction between the global North, where birth rates have largely plateaued, and the global South, where the lion's share of growth is occurring. This is inextricably linked to the relationship between family size, affluence and household security. Indeed, the population of Africa, for example, is predicted to double, from 1 billion to 2 billion, by 2050 (Foresight, 2011).

As Figure 2.2 also illustrates, we are currently in the midst of a fundamental shift in the population balance between rural and urban areas. It is estimated that in 2008 the numbers of people living in urban areas superseded those living in rural areas for the first time in recorded history (UNDESA, 2013). Moreover, urban population levels are predicted to continue to grow significantly over the next few decades, while global rural populations will actually decline. This rural-to-urban transition will occur in both the global North and South, though it is expected that the South will retain a larger percentage of the population in rural areas owing to the need for agricultural production capacity and corresponding lack of technology in agriculture among lower income nations. This shift is more than symbolic; it indicates an acceleration of human activity away from direct food production, and therefore a greater reliance on trade and global capital and therefore on currently long and complex supply chains. This potential distantiation, in terms of knowledge/skills and supply chain complexity, as well as spatially, could have a critical impact on food security, particularly when tied in with lack of affluence and associated health and environment issues.

Regarding the effect of global population numbers on the food system, population growth does not necessarily mean a proportional growth in food production requirements. It is well established that enough foodstuffs exist today to adequately feed the world's human population. The problem is not predominantly an issue of food production but rather an issue of food distribution and access.

Individual access to food, which has long been a problem, is still significant today and will in all likelihood intensify in the future. Moreover, the complexities of the food access problem have arguably increased over the past century as the food system has become increasingly globalized and dependent on market relations. There are now very few humans on the planet who are not dependent on the global food trading system, not least through its dominion over staples such as corn, maize and rice. Even localized food chains, which remain relatively common across the world, are impacted by the global system through competition for inputs (e.g. seeds, animal feed) and direct product competition. Indeed, the growth of local food systems in affluent Western markets over recent years can be interpreted as a direct consequence of global market conditions and consumer dissatisfaction with conventional products and markets. In this respect, this can be seen as an expression of a lack of resilience and therefore growing vulnerability in the dominant global food system.

In terms of human food producing capacity, the numbers of people directly engaged in agricultural production are thought to have risen very little over recent decades, while the non-agricultural population, mainly those populations concentrated in urban areas, according to FAO figures, have grown and will probably continue to do so (see Figure 2.3).

As Figure 2.4 illustrates, the majority of developed nations are experiencing a steady overall decline in the number of people who are directly reliant upon agriculture, hunting, fishing and forestry for their livelihoods. This is largely due to greater mechanization and yields in agriculture as well as increasing wealth and purchasing power among developed nation citizens (Wilson, 2002).

Broadly speaking, reaching a balance between food supply and food access requires reconciling the predominantly environmental, biological and socio-economic constraints that dictate food production with the predominantly political and socio-economic forces that create unequal access to food.

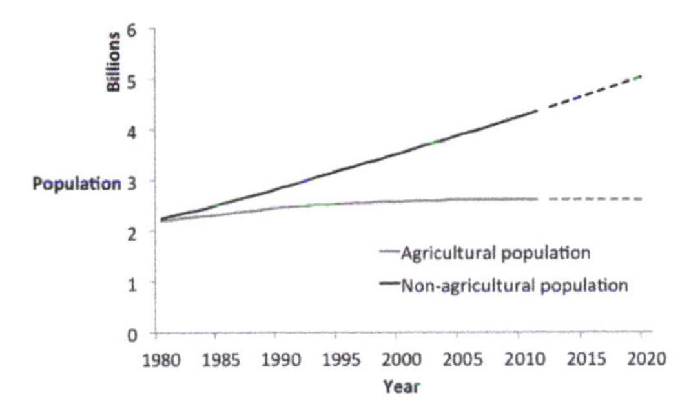

Figure 2.3 World agricultural population, 1980–2020.

Source: FAOSTAT population data.

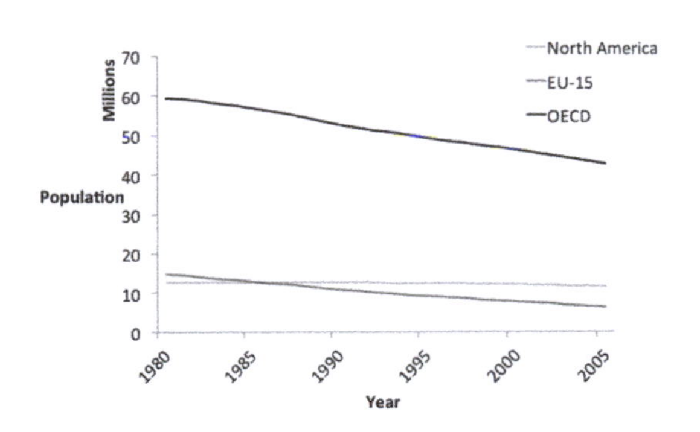

Figure 2.4 Economically active population in the agriculture, hunting, fishing and forestry industries, 1980–2005.

Source: Land Use for Agriculture, OECD, data compiled from FAO.

One highly significant potential impact of predicted global demographic changes over the coming decades is the effect of affluence on consumer demand. The 'nutrition transition' (Popkin, 2001), describes how typically, as countries develop and attain higher levels of affluence, their citizens use this wealth to acquire a more meat- and dairy-intensive diet. The adoption of these 'Western' consumption habits involves a transition away from predominantly plant-based foods, typically roots and tubers. The nature of this trend, and particularly its universality, are difficult to predict (Foresight, 2011) and would appear to depend on a deep interplay between sociocultural and bio-metabolic factors. Demand for animal-based products has significant impacts on both the environmental costs of the food system and potentially access to food for those less economically well off, as animals are a significant consumer of grain and soy. Over the last fifty years world average consumption of animal products has steadily risen (Figure 2.5), in line with general gains in affluence, and this trajectory is expected to continue.

Using data from the early 1960s and late 2000s, Figure 2.6 illustrates how animal-based food consumption has grown in rapidly developing countries compared to relatively stable consumption patterns in developed economies. Developing country demand for meat products nearly doubled between 1996 and 2007, with China accounting for the majority of the total increase in meat consumption (Ambler-Edwards *et al.*, 2009). Many predict a similar relationship between income growth and fruit and vegetable consumption, placing additional demands on resources such as land and water (IFPRI, 2012).

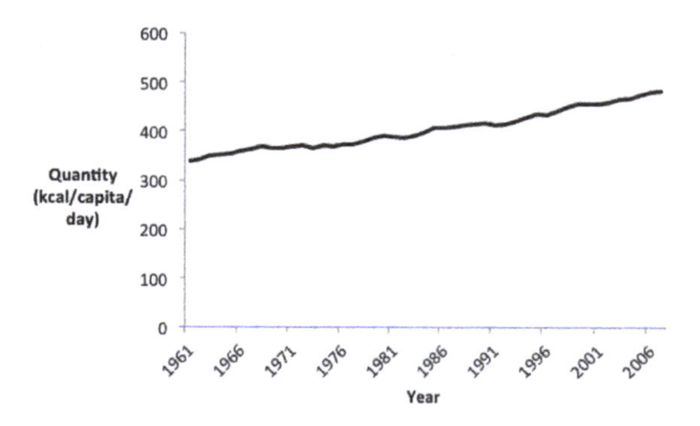

Figure 2.5 Global consumption of animal products, 1961–2007.

Source: FAOSTAT.

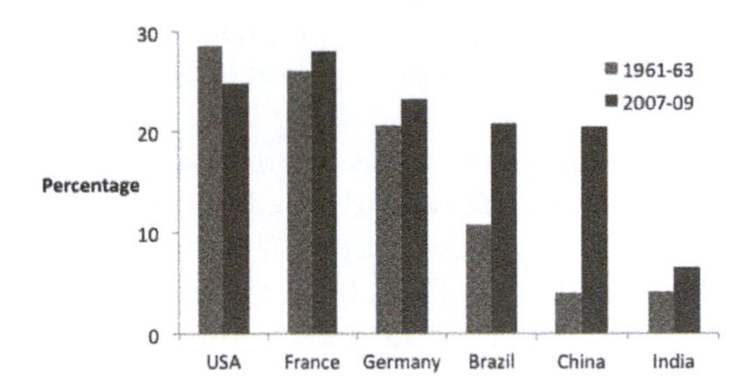

Figure 2.6 Share of animal-source foods in total dietary energy supplies, 1961–1963 and 2007–2009.

Source: FAO raw data (FAO, WFP and IFAD, 2012).

As mentioned above, existing animal production systems require high levels of grain and other plant-based inputs, the production of which both directly and indirectly impacts the price of grain and vegetable food stuffs in general. The effects of this are evidently most serious for those with limited economic power.

As Figure 2.7 illustrates, the global price of maize, soy and wheat increased significantly between 1998 and 2011. Apparent in this data are the price spikes, particularly post 2007. The overarching trend across this period appears to be an upward growth in prices. Many analysts predict that this growth in food commodity prices will accelerate in the short to medium term unless the underlying causes are addressed (see section 2, p. 48).

Long-term trends in human health and nutrition statistics indicate a gradual decrease in global levels of undernourishment. According to the FAO, 980 million people were thought to be undernourished – defined in terms of dietary energy supply – in 1990–1992, compared with 852 million in 2007–2009. This decrease, however, is thought to have levelled off since then, owing mainly to high food prices and economic problems, and remains significantly

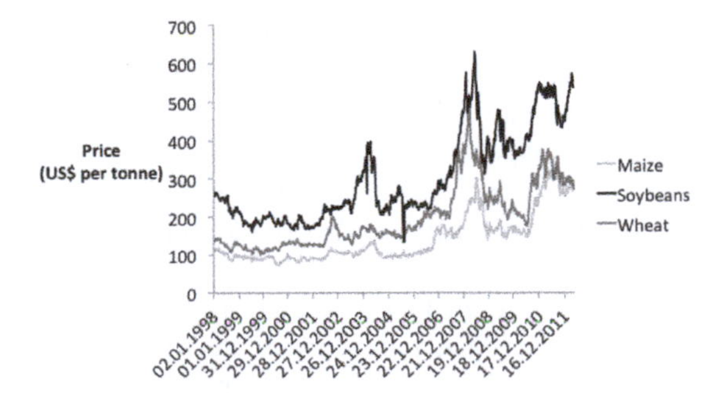

Figure 2.7 Global price of maize, soybeans and wheat, 1998–2011.

Source: FAO International Commodity Prices database.

higher than the Millennium Development Goal target of less than 500 million undernourished by 2015. A similar retardation of progress has been observed by the FAO in global poverty and child mortality rates (FAO, WFP and IFAD, 2012).

Increasing rates of overweight and obese populations are increasingly of concern in many Western developed countries, particularly the United States where 35.7 per cent of adults were classified as obese in 2009–2010 (up from a 22.9 per cent average between 1988 and 1994) (CDC, 2012). Cross-national comparisons are problematic owing to lack of data and methodological inconsistencies.

Environmental limits, pressures and vulnerabilities

Water

Average global water use between 1996 and 2005 has been estimated at 9087 Gm^3 per year. Of this, 92 per cent is used for agriculture and food production (Mekonnen and Hoekstra, 2011) with around 80 per cent being used for agricultural irrigation (IFPRI, 2002). Water availability is critically susceptible to potential impacts from climate change. The Intergovernmental Panel on Climate Change warned that by as soon as 2020, some African nations may witness decreases in yields of up to 50 per cent from rain-fed agriculture owing to climate change (IPCC, 2007). Water demand is also strongly linked to the nutrition transition through meat production requirements (Pimental and Pimental, 2003). The global trade in food means that approximately one fifth of global water use goes into food produced for export as embedded or virtual water (Mekonnen and Hoekstra, 2011).

Energy

In affluent countries, growth in per capita energy consumption for food production has consistently exceeded general energy consumption rates. Between 1997 and 2002, United States per capita energy usage declined 1.8 per cent while per capita food-related energy usage grew by 16.4 per cent. This is likely to be largely owing to the transition away from human labour to energy-intensive mechanized activities as well as greater demand for processed food products and increased transportation (Canning *et al.*, 2010). As Figure 2.8 illustrates, energy usage at the agricultural end of the food system has remained fairly constant over the past two decades among industrialized nations.

Fossil fuels are, of course, a vital resource for food production, whether it is for vehicle transportation or fertilizer production. Indeed, the rise in oil prices over recent years has been a significant factor in the rise in food prices. As Figure 2.9 illustrates, since the turn of this century, crude oil prices have risen significantly, particularly over recent years, and are widely predicted to continue to accelerate. Future oil prices will depend on the discovery of new reserves, extraction technology, geopolitics and global demand as well as the development of alternative energy sources. In the case of nitrogen fertilizer, the food system is currently reliant on natural gas as a raw material in order to produce enough food to meet demand. In common with the rest of the global economy, when and how the food system weans itself off its reliance on ultimately finite fossil fuels will be a key variable defining a transition to a more sustainable system.

Global agriculture's reliance on inorganic nitrogen fertilizer continues to grow, increasing by over 20 per cent between 2002 and 2009 alone (see Figure 2.10). Levels in North America and Europe, however, have remained fairly constant over this period, while consumption in China has grown significantly.

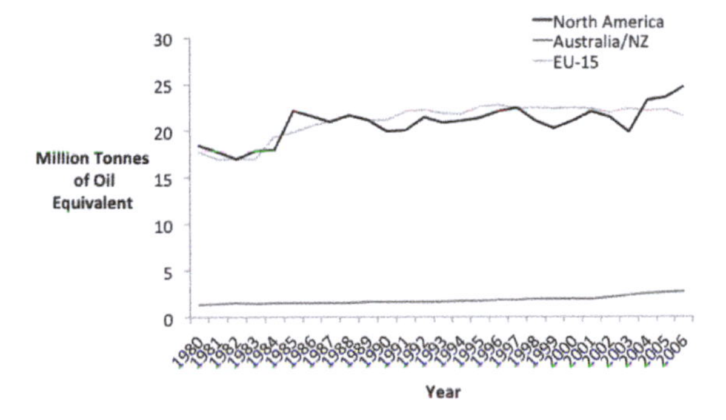

Figure 2.8 Total energy consumption by agriculture, 1980–2006.

Source: OECD–IEA.

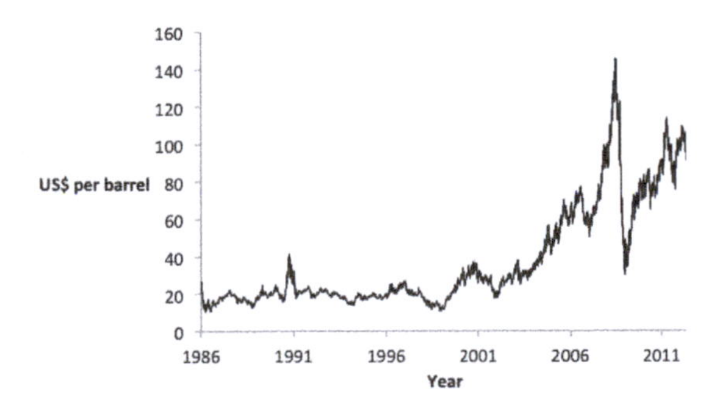

Figure 2.9 Crude oil prices, 1986–2012.

Source: US Energy Information Administration, Cushing, OK WTI Spot Price FOB.

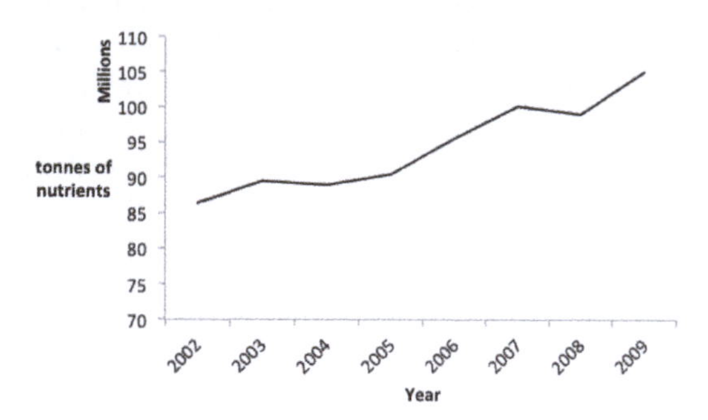

Figure 2.10 World consumption of nitrogen fertilizers, 2002–2009.

Source: FAOSTAT.

In recent years, the relationship between food production and energy has become further complicated by demand for biofuels. Since the mid 2000s, the production of bioethanol and biodiesel has grown significantly, largely as a reaction to high oil prices (see Figure 2.11), and government-led targets. During the 2007–2008 grain price spike it was estimated that 10 per cent of global course grain production was used for ethanol production (FAO, 2009). Although interest in this area is likely to have been tempered by high food prices, thus reducing the economic and strategic incentive to divert consumable resources into energy, it is probable that production will continue to increase, particularly as 'next generation technologies' come to market that use by-products and less directly competing resources. Public policy supports this drive, with US Law stating that 40 per cent of domestic corn production must be used for biofuel, while the EU retains a target of 10 per cent of all transport fuel being biofuel.

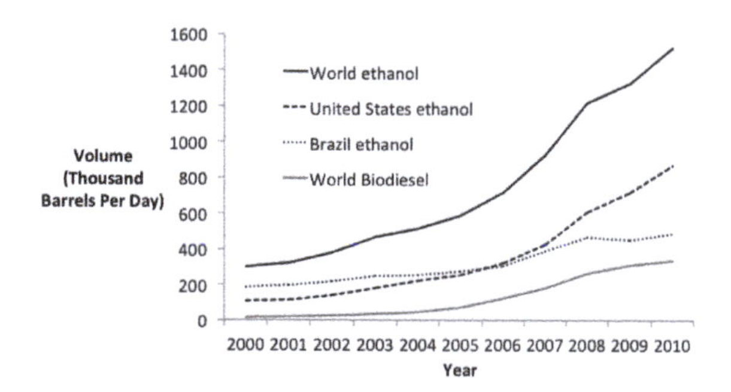

Figure 2.11 Ethanol and biodiesel production, 2000–2010.

Source: US Energy Information Administration.

Climate change

Climate change is evidently one of the major challenges for humanity over the next century. Efforts to mitigate anthropomorphic climate change are already well established, albeit with varying results. Globally, agriculture is thought to contribute around 14 per cent of all human greenhouse gas emissions as well as affecting the planet's ability to absorb or reflect heat owing to land use patterns.

As illustrated in Figure 2.12, EU nations have made steady progress in the reduction of greenhouse gas emissions from agriculture over recent years. Global emissions, however, continue to rise, from 4337 to 4689 million tonnes CO_2 equivalent between 2003 and 2010 (FAOSTAT, 2012).

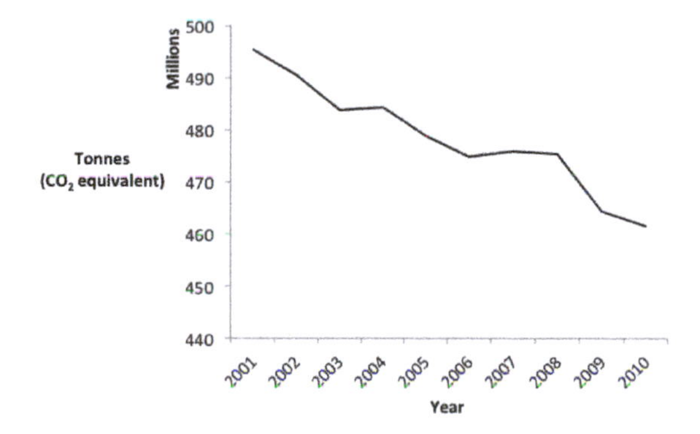

Figure 2.12 EU greenhouse gas emissions from agriculture, 2001–2010.

Source: Eurostat.

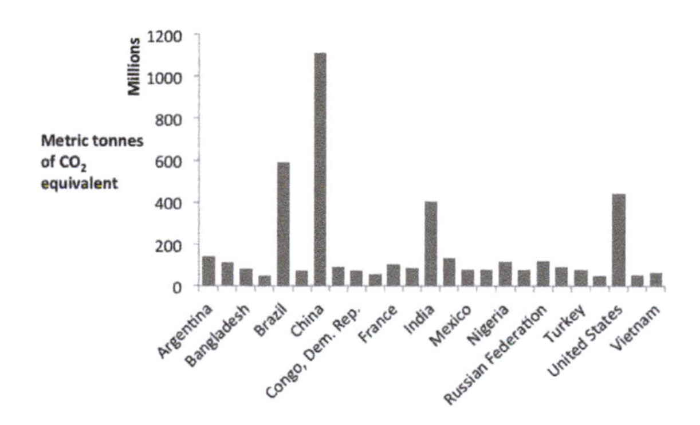

Figure 2.13 Non-CO_2 emissions from agriculture, 2005.

Source: The World Bank, World Development Report 2010: Development and Climate Change.

The impact of agriculture is particularly high regarding non-CO_2 gases, particularly nitrous oxide from soils and methane from animals. Globally, China, Brazil, India and the United States are by far the largest net emitters of non-CO_2 greenhouse gases.

It has been predicted that between 2000 and 2050 climate change will have an overall adverse impact on agriculture, with some countries and regions more significantly affected than others. As Table 2.1 sets out, the World Bank has made projections for individual countries over this period.

Table 2.1 Predicted impact of climate change on agriculture

	Projected physical impacts				*Projected agricultural impacts*	
	Change in temperature (°C)	*Change in heat wave duration (number of days)*	*Precipitation (% change)*	*Precipitation intensity (% change)*	*Agricultural output (% change)*	*Agricultural yield (% change)*
Country	*2000–2050*	*2000–2050*	*2000–2050*	*2000–2050*	*2000–2050*	*2000–2050*
Australia	1.5	10.9	–1.4	2.1	–26.6	–16.4
Bangladesh	1.4	8.7	1.4	5.4	–21.7	8.9
Brazil	1.5	13.5	–2.0	3.0	–16.9	–16.1
China	1.7	16.1	4.5	5.4	–7.2	8.4
Ethiopia	1.4	3.1	2.4	5.0	–31.3	0.5
Germany	1.5	14.8	2.4	5.0	–2.9	9.5
India	1.6	10.8	1.9	2.7	–38.1	–12.2
Japan	1.4	4.0	0.5	3.8	–5.7	0.6
Mexico	1.6	16.8	–7.2	1.6	–35.4	–0.5
Nigeria	1.3	4.1	0.6	1.1	–18.5	–9.9
Russian Federation	2.2	29.5	8.8	5.5	–7.7	11.0
Thailand	1.2	8.1	2.7	2.2	–26.2	–15.9
United Kingdom	1.1	5.1	2.5	3.7	–3.9	3.2
United States	1.8	24.4	2.7	4.0	–5.9	–1.7
Zambia	1.5	8.1	0.6	3.9	–39.6	1.3

Source: World Bank World Development Report 2010.

Among the most significantly affected nations in this model were India and Mexico, with predicted decreases of agricultural output of 38 per cent and 35 per cent respectively. The USA and Germany, by contrast, were projected to experience decreases of 6 per cent and 3 per cent respectively over this period.

Agricultural production

Although global food supplies are thought to consistently exceed demand in volume terms, distribution and unequal access to food remain persistent problems that make the assessment of both current and future food requirements more than an issue of global supply and storage. A 2012 FAO report estimated that approximately 10 per cent of the world's total potential caloric energy consumption is wasted, about 1.3 billion tons per year (FAO, 2012). Waste, inefficient resource use and diversion from human consumption are, of course, to some extent functions of market structure and its relationship to cost and purchasing power. Indeed under advanced capitalist conditions we know that the concentration of food companies and the increasing domination of markets by a small number of corporates has tended historically to favour overproduction, which in turn creates distorted under- and over-consumption problems. Equating a successful food system, in a productive sense, to globally balanced supply and demand is therefore a problematic concept, and one that is further undermined when considering environmental metrics for success.

Regarding food supply, agricultural land use is thought to constitute around 38 per cent of the Earth's surface (Foley *et al.*, 2011). A key statistic in terms of global agricultural production, however, is the overall decrease in agricultural land per capita, from 1.2 hectares in 1961 to 0.7 hectares in 2009 (see Figure 2.14).

Crop yields for corn, rice and wheat have risen steadily since the 1960s (see Figure 2.15), despite the amount of arable and permanent crop land remaining relatively static. The rate of growth in yields, however, appears to be in steady decline (FAO, 2009).

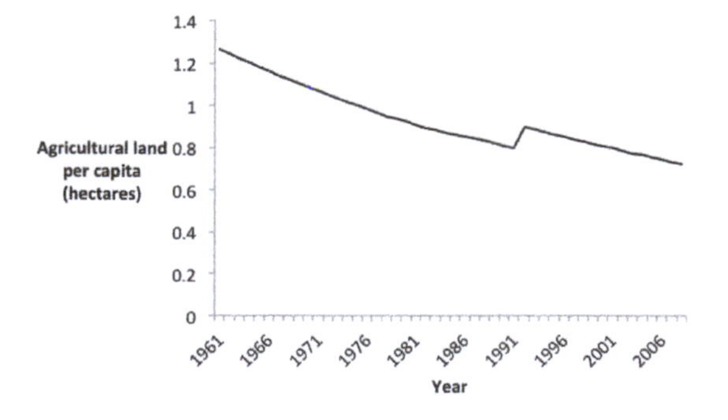

Figure 2.14 Global agricultural land per capita, 1961–2008.

Source: United Nations Environmental Programme EDE, United States Census Bureau International Programs Database.

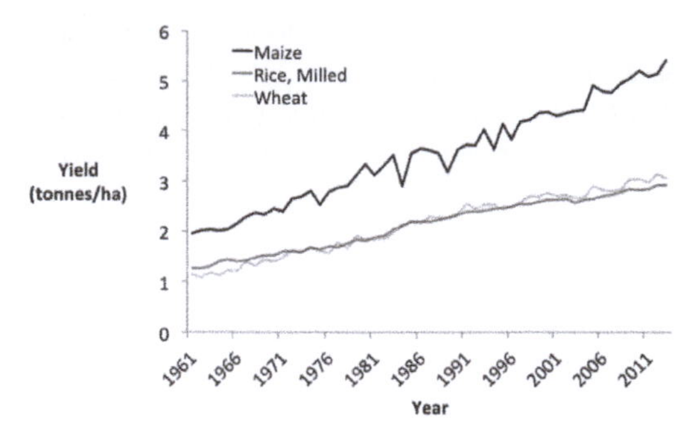

Figure 2.15 Global maize, rice and wheat yields, 1961–2012.

Source: Foreign Agricultural Service, Official USDA Estimates.

The storage of agricultural commodities has been an important mechanism of the modern food system to allow for production variation and therefore maintaining prices and availability. Global commodity stocks, however, have been at historically low levels during the past decade, owing to a combination of high food prices and political decisions, particularly by China and former Soviet states (Headey and Fan, 2010). Although these have recovered somewhat, largely owing to the impact of the 2007/2008 food crisis (see Figure 2.16) they remain at relatively low levels, leaving only a thin buffer of emergency supply when potential crises occur. Poor weather conditions in 2007, for instance, along with concurrent disease outbreaks in the beef, pork and poultry markets resulted in high food prices. While prices temporarily retracted from their peak in 2008/2009, they have since rebounded. According to IFPRI, this volatility is attributable to biofuel production, climate change and commodity future trading (IFPRI, 2011).

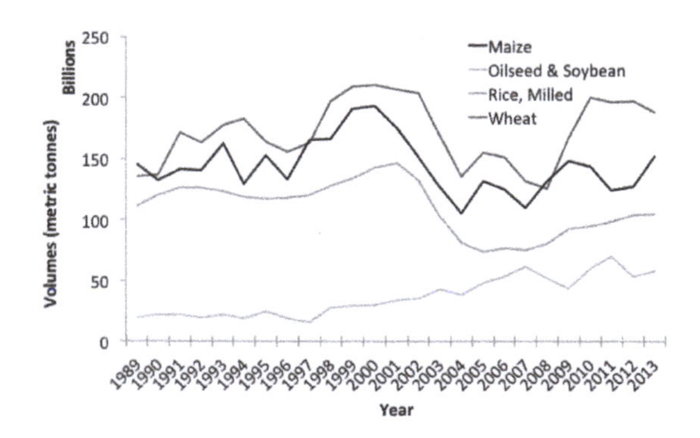

Figure 2.16 Global maize, wheat, oilseed and soybean, and rice stocks, 1989–2013.

Source: Foreign Agricultural Service, Official USDA Estimates.

There has been a general attempt to reduce direct state support to agriculture in recent years in order to open farmers to market pressures, decrease the taxpayer burden and meet commitments to global trade agreements. OECD figures reflect this, showing a consistent fall in support in the US, EU and across the OECD, when calculated as a proportion of total gross farm receipts (see Figure 2.17).

The adoption and use of transgenic crops continues to be one of the fastest growing segments in the agricultural industry with 160 million hectares being grown globally in 2011. Some twenty-nine countries, nineteen developing and ten industrial, now grow genetically modified crops. The US leads the way (see Figure 2.18) with 69 million planted acres, largely maize, cotton and soya. Brazil is rapidly developing in this area, however, with a growth rate of 20 per cent between 2010 and 2011 (ISAAA, 2011).

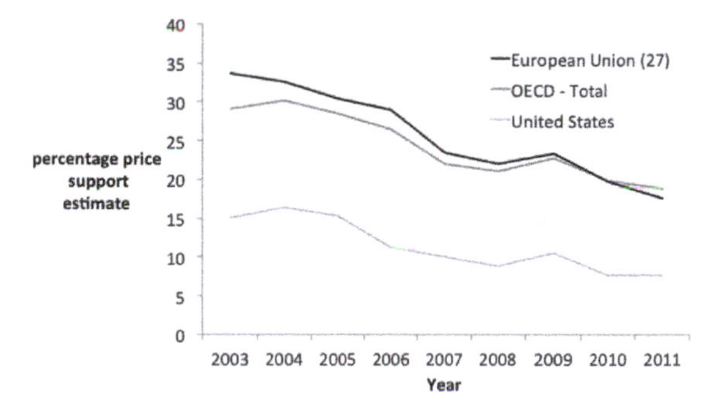

Figure 2.17 State support for agricultural producers, 2003–2011.

Source: OECD Agriculture Statistics.

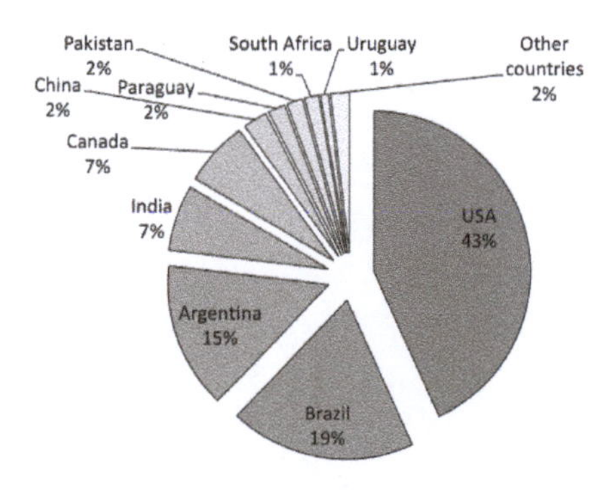

Figure 2.18 Global biotech crop production, 2011.

Source: ISAA Brief 43-2011.

Investment

According to the FAO, historically there has been under investment in agricultural public goods and services, despite their conclusion that it is one of the most effective ways to promote economic development. Indeed farmers themselves are identified as 'by far' the largest source of investment in agriculture, although this is largely in terms of capital stock rather than research and development (see Figure 2.19) (FAO, 2012b).

Foreign direct investment in agricultural production has grown significantly in recent years, particularly among World Bank classified Upper-Middle-Income Countries such as China, the Russian Federation and Brazil (see Figure 2.20).

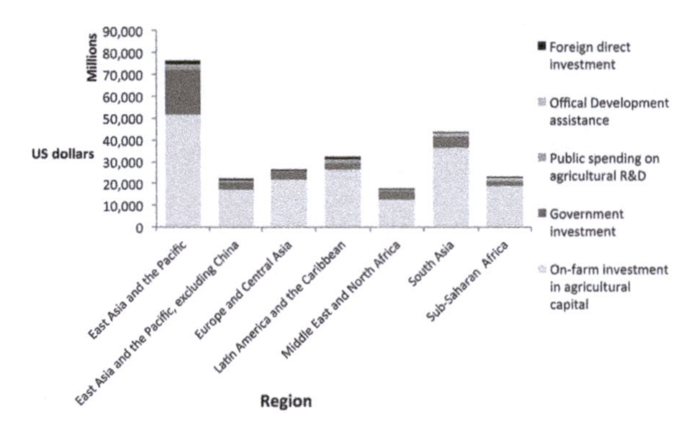

Figure 2.19 Agricultural investments in low- and middle-income countries.

Source: Lowder *et al.*, 2012.

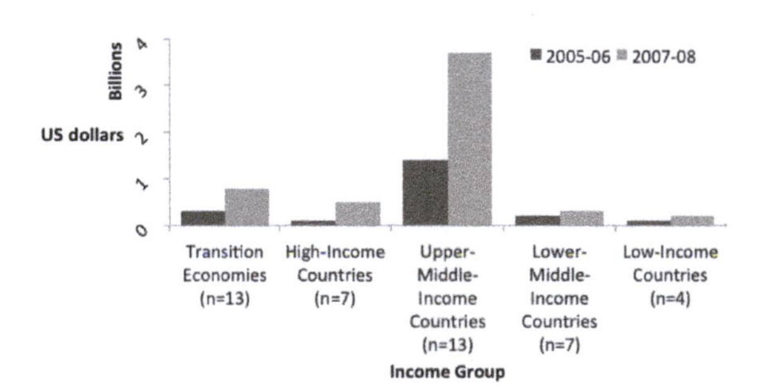

Figure 2.20 Average annual foreign direct investment in agriculture.

Source: FAO, *The State of Food and Agriculture, 2012.*

In some regions, however, state R&D expenditure has also increased significantly over recent years. Between 1997 and 2007, China, for example, increased public funding in this area from around $1 billion to $4 billion, while India more than doubled its public expenditure (FAO, 2012b). The United States is thought to invest around $10 billion per annum in total, with a roughly 50:50 split between public and private sources of funding.

Evidence of sustainability uptake

When seeking evidence for signs of a transition towards a more sustainable food system, we are unfortunately hampered by a lack of appropriate data sources at the global level. Organic agriculture is, of course, one of the clearest manifestations of a more sustainable approach to food production. Here we have clear evidence of an, albeit gradual, increase in land dedicated to a form of sustainable agriculture. Figure 2.21 shows increased organic production levels during the 2000s in both the EU and the United States. Globally, estimated certified production grew from around 6.5 million hectares in 2004 to 12.7 million hectares in 2009. Although this is encouraging, it still represents around 0.85 per cent of global agricultural land use (Willer and Kilcher, 2012).

Other proxies for sustainability indicate growth over recent years, although again, reliable aggregate data is uncommon. The market for certified fair trade products, for example, has risen consistently across Western industrialized nations. Sales grew by 12 per cent in the UK between 2010 and 2011, for example. Similarly, the popularity of 'local food' in countries of the global North has seen tremendous growth in the past two decades.

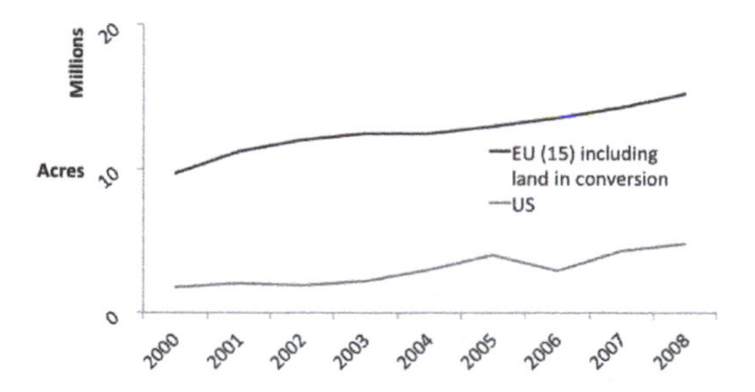

Figure 2.21 Area dedicated to organic agriculture, 2000–2008.

Source: USDA ERS, Organic Europe.

For instance, in the United States, direct-to-consumer sales rose $649 million between 1997 and 2007 while the number of farmers' markets rose nearly 200 per cent between 1994 and 2009 (USDA, 2010).

The difficulties associated with quantifying the uptake of sustainable food can be attributed to three factors. First, sustainable food is a contested and dynamic term. Second, in practice sustainable food systems are typically small scale and diverse (Horlings and Marsden, 2011). Third, as a nascent and relatively niche aspect of the food system, there appears to be a lack of interest from public funders to measure its impact in quantitative terms, particularly at a global level. A rich and growing body of evidence exists on individual initiatives and small-scale systems, usually case study based and reliant on qualitative accounts. Quantifying the scale and scope of sustainable food initiatives across larger realms, however, remains an important task for sustainability advocates.

The foregoing analysis then has identified the most significant reasons for the growing vulnerability, insecurity and unsustainability of our present global food system. Increasing food access issues have become compounded by systemic environmental limits and resource deficiencies and scarcities. These trends have added to the volatility of food 'markets', and created the spectre of food supply shortages, even within globally organized retail-led supply chains. With low levels of food stocks, the current system overproduces and is increasingly wasteful on the one hand, and through its supply chains and trading policies creates growing scarcities on the other. Moreover this 'just-in time' system is based upon a per capita shrinking of the productive land at the same time as more competition for this resource by the energy and bioeconomy sectors. The result is a system under stress with a series of vulnerabilities in supply and a variety of social and political crises. The following section focuses on the institutional responses to this 'perfect storm'.

Section 2: Perspectives: towards sustainable food futures

One of the major impacts of the 2007/2008 commodity price spike was to focus attention on the broader sustainability of the food system. For many, it was a first manifestation of the growing set of pressures on the system. Moreover, the impact on consumer prices, which brought increased hardship to millions of impoverished consumers around the world and sparked social unrest (Headey and Fan, 2010), demonstrated a clear link between environmental pressures and the ability of the system to produce enough food.

Among the major political and institutional responses to the crisis was the development of a series of future prediction exercises that aimed to provide a strategic 'steering' for food related policy and institutional behaviour. Although terms of references varied, including temporal horizons and substantive priorities, these documents share a common conceptual basis in that they explicitly link the twin challenges of environmental sustainability and food security.

This section identifies some of these key documents and attempts to both summarize dominant themes and draw parallels between the key challenges identified, the nature of solutions proposed and the perspectives of the stakeholders involved in their authorship. In this respect, of interest here is both the identification of possible strategies to both mitigate against and adapt to the impacts of the growing resource pressures outlined in the first section, and an analysis of the process of developing these responses and in particular how the crisis is framed and by whom.

International governmental organizations, particularly the FAO and other UN bodies, have led the production of these documents. National governmental responses have also been produced as well as some by NGOs and academics. Significantly, a small number of 'business' responses to the food sustainability crisis have been produced. As a starting point, Table 2.2 lists nine of the most significant institutional reports over recent years and summarizes their key approaches to the issue.

In addition, a number of important academic-based publications that consider the future of food have been produced in recent years. These include broad participatory scoping exercises (Ambler-Edwards *et al.*, 2009), large multi-authored approaches to food security (Godfray *et al.*, 2010) and agricultural development (Pretty *et al.*, 2010; Foley *et al.*, 2011) and sector specific responses (Jaggard *et al.*, 2010).

Although the framing adopted by these publications is clearly influenced by the aims of the specific objects of the processes underpinning their production, their institutional context and when they were produced, both common features and some distinctive approaches can be identified. The studies consistently recognize the interconnectivity of food by adopting a holistic global perspective, even if only to aid the formulation of a geographical or sectorial response, and connecting meta-themes of food security and environmental sustainability.

Table 2.2 Recent food futures reports and their key focal points

Title	*Year*	*Key focal points*
How to Feed the World in 2050 FAO High Level Expert Forum	2009	The required production/technology improvements and social/political reforms necessary to ensure access to food. Topics include public/private investment in developing nation agriculture, climate change impacts, and the importance of efficient global trade.
Towards the Future We Want FAO	2012	Presented three primary messages for the 2012 Rio+20 Summit: the eradication of hunger and malnutrition, improvements in consumption and production systems, and reforms in food governance.
Future of Food and Farming The UK Government Office for Science	2011	Focuses on how to balance competing pressures and demands on the future food system. The key challenges identified were: balancing supply and demand, maintaining biodiversity and ecosystem services, moving to a low emission food system, addressing future volatility and eliminating hunger.
Vision 2050 World Business Council for Sustainable Development	2010	Presents business perspectives that focus on how to achieve a sustainable future and the role of being able to make sustainable choices in a marketplace environment.
Scaling up in Agriculture, Rural Development, and Nutrition International Food Policy Research Institute	2012	A compilation of articles broadly focused on the scaling up of agriculture, rural development and nutrition. The key topics are: role of rural community engagement, value chains, nutrition interventions, institutional approaches and international aid.
Achieving Food Security in the Face of Climate Change Beddington *et al.*, for CGIAR	2011	Proposes evidenced-based actions to achieve food security including: integration of food security and sustainable agriculture into global and national policy, significant increases in global investments in sustainable food systems, an emphasis on sustainable intensification, a focus on those most vulnerable to food insecurity and climate change, and reductions in food waste.
Scaling up Global Food Security and Sustainable Agriculture UN Global Compact	2012	Highlights 'leading edge' practices in the private sector to encourage progress in food security and sustainable agriculture. Presents five 'pillars' of food security and sustainable agriculture: sustainable sourcing, improvement of land and water management, enhancing nutrition, effective use of technology, reductions in commodity price volatility.
Environmental Sustainability Vision Towards 2030 FoodDrink Europe	2012	Reviews key actions needed to achieve environmental sustainability by 2030. These include more sustainable sourcing of agricultural products, resource efficiency, and sustainable consumption and production. Promotes industry action in achieving European environmental sustainability.
Food and Agriculture: the Future of Sustainability UN Department of Economic and Social Affairs	2012	Aims to illustrate how 'leading thinkers imagine our future food and agricultural world'. The conclusions include a focusing on small and medium farmers, working towards human nutrition as the primary goal, reducing waste and inefficient use of resources and measuring impacts.

Food production and security

The major governmental reports adopt a relatively optimistic approach to this theme, provided that the changes they put forward are implemented. For example, the FAO High Level Expert Forum report *How to Feed the World in 2050*, which focuses on food security, predicts being able to meet the food demands of 2050 as long as effective investment and policies are put in place. They estimate that approximately $89 billion needs to be invested in developing world agriculture alone over the preceding period. Around $50 billion of this figure is required for 'downstream' services, incorporating marketing and supply chain infrastructure and including storage facilities. The issue of which constituents on the food system will be required to make the necessary investments is highlighted here by their conclusion that most of this investment should come from private investors, including farmers and supply chain businesses (FAO, 2009).

Addressing 'inefficiency' in agriculture, particularly within the developing world, is frequently identified as a vital element. In an article published in *Science*, Godfray *et al.* (2010) stress the importance of reducing the 'global yield gap' between strong and weak performers, much of which, in their view, can be achieved through education and skill improvement, and without recourse to fundamental changes in farming systems and the widespread use of biotechnology. If achievable yield improvements can be made, the FAO estimate that only 20 per cent of the required increase in production would need to be from expanding agricultural land use (FAO, 2009).

Reducing trade barriers and other market distortions is another common action proposed in food futures reports. Open and 'fair' global trade is recognized as important, especially for developing nations where open market access is regarded as crucial to minimize market distortions (FAO, 2009). The continued decoupling of price support mechanisms, along with increased direct payments for environmental services, is put forward as an important element in this regard, particularly within the EU. According to the UK government *Future of Food and Farming* report, rather than emphasizing the need for national self-sufficiency, rejected as being inefficient, the world would benefit from adopting a global vision of food security based on cooperation and trade (Foresight, 2011). In this way, the authors link agricultural production to food access issues and argue for the removal of political and economic obstacles to market forces.

Focusing on historic growth rates in global food production, an FAO study observes that the average rate increase of 2.2 per cent over the past decade is about double what is required to sustain a 60 per cent higher demand for food in 2050 (FAO, 2011). The key factor therefore, under the assumption that this rate can be maintained, would appear to be not whether human ingenuity can meet the output challenge but whether it can be done in a way that respects environmental limits and feeds the global poor (FAO, 2012a).

On the question of transgenic technology, the major FAO food futures reports are non-committal. The UK *Future of Food and Farming* report, however, outlines the importance of ensuring low-income country access to 'new technologies such as GM to enhance traits' (p. 88) such as drought tolerance. The report also identifies the need to encourage consumers and food system constituents to overcome ideological barriers to embracing GM technologies.

According to the FAO's Rio+20 report, *Towards the Future We Want*, however, global food security can be achieved without having to match net supply and demand growth, through innovative reforms in the governance of agricultural and food systems (FAO, 2012a). The emphasis for reform, according to this document, should be on measures that work towards the right to food and equality. In this respect, food security is framed as being as much about effective governance as it is about improving crop yields, therefore implying a strong and more explicit link between food security and sustainability.

The importance of small-scale farming is picked up in some of the food futures reports. The FAO's Rio+20 submission report, for example, emphasizes the value of smallholders in both a development and food security context. Agricultural extension and appropriate infrastructure are among the key factors for supporting these sectors (FAO, 2012a). The same principles apply, albeit often with differing terminology, in the developed world context with recognition of the rural development benefits of small food producers and their potential in meeting local demand (Foresight, 2011). The earlier FAO document *How to Feed the World in 2050* supports the observation that economic growth strategies should focus on agricultural development and smallholders, which have tended to be overlooked over recent decades (FAO, 2009). Studies have shown that investments in these areas are at least twice as effective in benefiting the poor as general economic growth (World Bank, 2008). Fostering general economic development in the developing world is also identified as a way to promote overall food security, despite problems associated with the nutrition transition.

Environmental sustainability

Achieving environmental sustainability is unanimously regarded as a central challenge for the food system. The dominant approach to achieve this centres largely on forms of 'sustainable intensification', and in particular closing the productivity 'yield gap' in a way that sustains acceptable levels of biodiversity, ecosystem services and other environmental resources (Foley *et al.*, 2011). Godfray *et al.* (2010) promote strategies such as precision agriculture, agroforestry, and integrated pest management as a means to improve yields while minimizing the environmental burden and the need for agricultural land use expansion in particular. Jaggard *et al.* (2010) estimate that at least a 50 per cent increase in crop production could occur if additional intensification measures were taken, including technologies and methods such as transgenics and innovative plant breeding.

Food waste and the need to reduce the amount of food that fails to be consumed is a recurrent action point. Some estimates put the proportion of food lost in this way at between 30 and 50 per cent of total global food production (IMECHE, 2013). The *Future of Food and Farming* report proposes reducing food waste by half by 2050, although it acknowledges difficulties associated with defining what waste is, its relation to inefficient resource use and the difficulties associated with recording its occurrence accurately (Foresight, 2011).

Climate change is, of course, a central concern and one that is acknowledged to add a high degree of complexity to future prediction exercises, in terms of both its role in mitigation and potential adaptation responses. A prominent report focusing on the impact of climate change on the food system is *Achieving Food Security in the Face of Climate Change* (Beddington *et al.*, 2011), published by the Consultative Group on International Agricultural Research (CGIAR). Among the strategies put forward to mitigate adverse impacts, particularly on the poor, are recommendations for elevated rates of investment in sustainable agriculture, policies focused on food insecurity, and development of human–ecological information systems that permit regular measurement and monitoring of these systems.

Business perspectives

Business interests often feed into food future reports, usually through stakeholder consultation processes (see Ambler-Edwards *et al.*, 2009; EFRAC, 2009). How much these reports reflect the perspectives and objectives of business stakeholders appears to depend on a multitude of factors, not least the objectives of the sponsoring organization and the political philosophy of the associated governing officials. There are, however, a number of reports led by business that perhaps provide a less adulterated picture of commercial perspectives on the sustainability challenge.

The *Environmental Sustainability Vision Towards 2030* report by FoodDrink Europe, an industry association of European food and drink companies, is a report focused on environmental sustainability in general and the leadership role private business can take in achieving a vision of sustainability set out in the report (2012). The document presents a series of case study examples of sustainable practice, such as coffee sourcing by Ferrero and Unilever's pledge to source all of its agricultural raw materials from sustainable sources by 2020.

Vision 2050, produced by the World Business Council for Sustainable Development in 2010, is a major report on the challenge of sustainability in general (WBCSD, 2010). The document promotes the ability of the market to account for negative externalities associated with the business sector, including food production. Among the issues put forward is the need for increased productivity through research and development investments. The report, whose signatories include PricewaterhouseCoopers, The Procter & Gamble Company and Toyota Motor Corporation, considers the role business can play in achieving a sustainable world by 2050. It emphasizes the commercial benefit of an increased number of consumers, but also the limiting effects that climate change and resource depletion may have on people's ability to 'maintain the consumption lifestyle that is commensurate with wealth in today's affluent markets' (WBCSD, 2010, p. 3). In terms of the role of agriculture in a sustainable 2050, Table 2.3 sets out a pathway based on a sustainable intensification model with technological solutions and productivity gains at its heart.

In a discussion of the most pressing challenges facing the private sector's involvement in food security and sustainable agriculture, the UN Global Compact report *Scaling up Global Food Security and Sustainable Agriculture* highlights the importance of the food and agriculture industry in economic terms, accounting for around 10 per cent of global GDP (UN Global Compact, 2012). This industry-focused report presents the following five pillars of food security and sustainable agriculture:

- sustainable sourcing, including responsible sourcing from SMEs, and the development of working definitions of sustainability through processes such as certification schemes and industrial cooperation;
- improving land and water management, through stronger collaboration and innovation to ensure effective management while growing sustainable agricultural production;
- enhancing nutrition, through private and public sector approaches to providing affordable and nutritious food;
- effectively using technology, and recognizing its importance as a tool to improve food security and enhance sustainable agriculture;
- reducing commodity price volatility through public and private sector initiatives to aid the poorest and most vulnerable populations.

Table 2.3 Pathway to 2050: agriculture

Our vision	Enough food and biofuels through a new Green Revolution
Measures of success	Agricultural output doubled by improved land and water productivity
'Must haves' by 2020	Training of farmers Freer and fairer trade Yield gains Water efficiency More agri R&D New crop varieties
Key themes for 'turbulent teens' (defined as 2010–2020)	Cultivating knowledge intensive agriculture
Key themes for 'transformation times' (defined as 2020–2050)	Growth in global trade, crop yield and carbon management

For each pillar the report provides a series of company case studies that showcase examples of private sector success. More broadly, the report frames the scaling up of sustainability as a consumer-driven paradigm change.

A UK scenario-based consultation with industry stakeholders reported a consensus that government should lead a transition to a more sustainable 2025 through appropriate regulation and legislation (Livesey *et al.*, 2010). Through industry-focused consultations, the research recommended the following series of 'potential actions':

- a clear and shared vision across stakeholders;
- deeper and more open consultation;
- stronger evidence for the measures that are required to meet the goals we set;
- consistent and coherent regulation;
- changing consumer expectations;
- increasing the skills base of the food industry;
- continuing to focus on innovation with clear sustainability targets;
- embracing change.

This desire that government leads the challenge was also stated during the consultation behind the UK Foresight *Future of Food* exercise (GOS, 2011).

Another significant business-centred food futures report is *Food and Agriculture: The future of sustainability* published by the UN Department of Economic and Social Affairs (Giovannucci, 2012). Through consultation with four distinct interest groups (Policy; Rural Livelihoods and Poverty; Agricultural Production and Environmental Sustainability; and Business Specialists), the authors report the following business- and technology-focused issues where consensus within and between the groups was hard to reach:

- Will large- or small-scale farming best deliver food security?
- What roles should corporations have in our food systems?
- What agricultural production technologies will best deliver sustainable food security?
- What role could GM play in improving food security?
- How much agro-biodiversity should we promote in our farming systems?
- How can we adapt to growing demand for livestock products?
- How can trade best affect food security?

Together, these business-focused reports clearly indicate a distinct set of motivations and proposed solutions compared to the major largely governmental or academic documents. They may be characterized as promoting a market-based transition to a sustainable food system based on informed consumer choice and 'enabling' policies by government. Although the activities of organizations such as the World Business Council for Sustainable Development have been criticized as greenwash (Najam, 1999), these documents articulate the perspectives of a major part of the food system. How proactive the business community will be remains to be seen, of course.

Alternative futures?

Many reports by governmental bodies include what may be considered as 'alternative' practices as part of future strategies and responses to sustainability pressures. Indeed these documents tend to at least construct a marginal space for the framing of alternative agri-food systems such as local food, organic production and community supported food initiatives. On the whole, however, these initiatives are considered as part of the solution, coexisting with a dominant conventional sector. This is particularly the case for reports that focus on food security and supply rather than environmental security. A number of NGO advocacy groups, however, have presented their own future visions of a food system with what can be considered alternative practices at their heart. The UK Soil Association, for example, have produced a series of reports that present the case for a more localized, health-based food system based on organic production principles (Soil Association 2008, 2009, 2012). The Californian think tank Institute for the Future suggest a series of key shifts in the food system that include decentralization and the development of more 'personal' relationships with food by consumers (Institute for the Future, 2010).

The international NGO Oxfam published its *Growing a Better Future* report in 2011 which outlined three key challenges that the food system must meet: the sustainable production challenge, the equity challenge and the resilience challenge. The equity challenge is the main point of departure from the other reports documented in this chapter, stressing the need to empower the grassroots poor, particularly smallholders. The report heralds positive developments already underway and calls for new global governance structures in order to support the required transformation that must come from both the top down and the bottom up (Bailey, 2011).

Section 3: Implications for building sustainable food futures: a conclusion

As Section 1 illustrates, there can be no reasonable doubt that the global system of food provision is under a serious and growing set of environmental, social and economic pressures that threaten its long-term sustainability.

Despite a necessity for caution regarding data provision and reliability, and the need to acknowledge difficulties associated with predicting future developments, it is clear that current practices must change in order to 'feed the planet' in a manner that does not compromise the ability of future generations to do the same. Whether unforeseen technological or social developments or shocks to the system originating from nature will prove to alter the course of events can only be speculated upon. There is a growing imperative, however, to develop an understanding of the threats to the system as they currently stand and aid the strategic development of responses to this crisis at multiple levels including government, business, civil society and communities.

As the second section of this chapter illustrates, contemporary strategic responses to the food crisis, albeit those limited to official, written and publicly available documents, demonstrate a largely common framing of the issues with a degree of dissonance regarding response priorities. The content and overall approach of these documents clearly depend largely on the nature of the organizations that have driven their development, what their terms of reference were, and when the documents were produced. It is also clear, however, that the dominant framing of responses within these documents is one of system evolution, rather than system change, and a largely neoliberal philosophy regarding the stimulation of appropriate responses. In this respect, these responses would appear to maintain the hegemonic status of free markets, based on global standardization and minimal public regulation. Despite growing popular interest in ecological and social responses to food sustainability issues at both local and non-local levels, 'alternative' responses risk remaining marginalized by these dominant discourses that prioritize standardized, techno-focused, market-based responses.

As concluded by the Foresight report (2011), addressing the challenges of food futures will necessitate integrated approaches to policy and governmental decision making that have thus far been difficult to reach. The breadth and depth of this issue dictates that diverse policy areas that have traditionally worked largely in isolation must come together. This principle also applies vertically, as the various levels of subnational, national and super-national government must work together to ensure effective multilevel governance. It also applies between government and other key constituents of the food system, most notably the business sector and advocates of alternative approaches. Effective solutions will inevitably involve these respective approaches trading off their core interests for the good of the common food system.

Questions should be asked of course, regarding the power of these strategic documents to effect change, both within government and business. As set out in the introductory chapter of this book, short-termism is rife within sustainable policy making and one of the key barriers to the development of a sustainable food system. Policy makers must ensure that these documents are not left to 'gather dust'. Strategies must be reappraised, on the basis of robust data, and effective solutions developed that account for the true holders of power (both human and non-human) within the food system. This is a clear role for social scientists. Where power to effect change is currently out of reach, those charged by society as guardians of our food system must acquire the means to do so, whether through market regulation, investment or education.

The following chapters in this book aim to aid this process by focusing on some of the key aspects of the food system, and through their analysis point to some of the important opportunities and constraints. Two vital dimensions highlighted both within this chapter and those that follow are the degree of largely inherent complexity and the levels of both conceptual and institutional embeddedness retained within the current system that may retard positive change.

It remains to be seen whether the proliferation of food futures reports described in the second part of this chapter will continue. Whether they continue to be produced will in some ways be a test of their strength as drivers of change. What they have achieved to date, however, is to help re-establish concern over food scarcity and its associated dimensions of sustainability. In general, this wave of food futures documents can be regarded as largely promoting a case for forms of 'sustainable intensification' and calling for stakeholders across the food system to work towards effective manifestations of this concept. At the same time, however, it should be recognized that these dominating framings of the food crisis have the effect of marginalizing alternative views not based on sustainable intensification. Neo-Malthusians, deep-greens and techno-optimists all have a role in working towards a sustainable food system. As outlined at the beginning of this chapter, history demonstrates that humans consistently overestimate their ability to predict future developments. The strategic development of a sustainable food system must therefore be achieved through reflexive governance systems built on sound data and science.

References

Ambler-Edwards, S., Bailey, K., Kiff, A., Lang, T., Lee, R., Marsden, T., Simons, D. and Tibbs, H. (2009) *Food Futures: Rethinking UK strategy*, London: Chatham House.

Bailey, R. (2011) *Growing a Better Future: Food justice in a resource constrained world*, Oxford: Oxfam International.

Beddington, J., Asaduzzaman, M., Fernandez, A., Clark, M., Guillou, M., Jahn, M., Erda, L., Mamo, T., Van Bo, N., Nobre, C.A., Scholes, R., Sharma, R. and Wakhungu, J. (2011) *Achieving Food Security in the Face of Climate Change: Summary for policy makers from the Commission on Sustainable Agriculture and Climate Change*, Copenhagen: CGIAR Research Program on Climate Change, Agriculture and Food Security (CCAFS).

Canning, P., Ainsley, C., Huang, S., Polenske, K.R. and Waters, A. (2010) *Energy Use in the U.S. Food System*, Washington DC: U.S. Department of Agriculture, Economic Research Service.

CDC (2012) 'NCHS Health E-Stat. Prevalence of overweight, obesity, and extreme obesity among adults: United States, trends 1960–1962 through 2009–2010', Center for Disease Control, available at: www.cdc.gov/nchs/data/hestat/obesity_adult_09_10/obesity_adult_09_10.htm, accessed 13 June 2013.

Compton, J., Wiggins, S. and Keats, S. (2010) *Impact of the Global Food Crisis on the Poor: What is the evidence?* London: Overseas Development Institute.

EFRAC (2009) *Securing Food Supplies up to 2050: The challenges faced by the UK*, Fourth Report of Session 2008–2009, vol. I, House of Commons Environment, Food and Rural Affairs Committee, London: HSMO.

FAO (2009) *How to Feed the World in 2050*, FAO High Level Expert Forum, Rome: FAO.

FAO (2011) *Global Food Losses and Food Waste, Extent, Causes and Prevention*, by J. Gustavsson, C. Cederberg, U. Sonesson (Swedish Institute for Food and Biotechnology) and R. van Otterdijk and A. Meybeck (FAO), Rome: FAO.

FAO (2012a) *Towards the Future We Want: End hunger and make the transition to sustainable agriculture and food systems*, FAO at Rio+20, Rome: FAO.

FAO (2012b) *The State of Food and Agriculture 2012: Investing in agriculture for a better future*, Rome: FAO.

FAO (2012c) 'FAO food outlook, May 2012', available at: www.fao.org/giews/english/fo/index.htm, accessed 13 June 2013.

FAO, WFP and IFAD (2012) *The State of Food Insecurity in the World 2012: Economic growth is necessary but not sufficient to accelerate reduction of hunger and malnutrition*, Rome: FAO.

FAOSTAT (2012) FAOSTAT emissions database, www.faostat.fao.org, accessed 08 December 2012.

FAOSTAT (2013) FAOSTAT food aid shipments database, www.faostat.fao.org, accessed 17 January 2013.

Foley, J.A., Ramankutty, N., Brauman, K.A., Cassidy, E.S., Gerber, J.S., Johnston, M., Mueller, N.D., O'Connell, C., Ray, D.K., West, P.C., Balzer, C., Bennett, E.M., Carpenter, S.R., Hill, J., Monfreda, C., Polasky, S., Rockström, J., Sheehan, J., Siebert, S., Tilman, D. and Zaks, D.P.M. (2011) 'Solutions for a cultivated planet', *Nature*, vol. 478, no. 7369, pp. 337–342.

FoodDrink Europe (2012) *Environmental Sustainability Vision towards 2030: Achievements, challenges and opportunities*, Brussels: FoodDrink Europe.

Foresight (2011) *The Future of Food and Farming: Final project report*, The Government Office for Science, London: HMSO.

Giovannucci, D., Scherr, S., Nierenberg, D., Hebebrand, C., Shapiro, J., Milder, J. and Wheeler, K. (2012) *Food and Agriculture: The future of sustainability: A strategic input to the sustainable development in the 21st century (SD21) project*, New York: United Nations Department of Economic and Social Affairs, Division for Sustainable Development.

Godfray, H.C.J., Beddington, J.R., Crute, I.R., Haddad, L., Lawrence, D., Muir, J.F., Pretty, J., Robinson, S., Thomas, S.M. and Toulmin, C. (2010) 'Food security: The challenge of feeding 9 billion people', *Science*, vol. 327, no. 5967, pp. 812–818.

GOS (2011) *Developments in the Global Food Supply Chain: Summary of workshop and key messages*, Foresight, The Government Office for Science, London: HMSO.

Headey, D. and Fan, S. (2010) *Reflections on the Global Food Crisis: How did it happen? How has it hurt? And how can we prevent the next one?* Washington, DC: IFPRI Research Monograph, International Food Policy Research Institute.

Horlings, L. and Marsden, T. (2011) 'Towards a real green revolution: Exploring the conceptual dimensions of a new ecological modernization of agriculture that could feed the world', *Global Environmental Change*, vol. 21, no. 2, pp. 441–452.

IFPRI (2002) *World Water and Food to 2025: Dealing with scarcity*, Washington DC: International Food Policy Research Institute.

IFPRI (2011) *Global Food Policy Report, Food Prices: Riding the rollercoaster*, Washington, DC: International Food Policy Research Institute.

IFPRI (2012) 'Ensuring food and nutrition security in a green economy', IFPRI Policy Brief 21, June 2012, Washington DC: International Food Policy Research Institute.

IMECHE (2013) *Global Food: Waste not, want not*, London: Institution of Mechanical Engineers.

IMF (2012) *World Economic Outlook*, April, Washington, DC: International Monetary Fund.

Institute for the Future (2010) *Food Web 2020*, Palo Alto, CA: Institute for the Future.

IPCC (2007) *Climate Change 2007: Synthesis report. Contribution of Working Groups I, II and III to the Fourth Assessment Report of the Intergovernmental Panel on Climate Change*, Geneva: Intergovernmental Panel on Climate Change

ISAAA (2011) *ISAAA Brief 43-2011: Highlights: Global status of commercialized biotech/GM crops: 2011*, New York: International Service for the Acquisition of Agri-biotech Applications.

Jaggard, K.W., Qi, A. and Ober, S. (2010) 'Possible changes to arable crop yields by 2050', *Philosophical Transactions of the Royal Society B: Biological Sciences*, vol. 365, no. 1554, pp. 2835–2851.

Livesey, F., Frau, I., Oughton, D. and Featherstone, C. (2010) *Future Scenarios for the UK Food and Drink Industry: Report to the Food and Drink Federation*, Institute for Manufacturing, Cambridge: University of Cambridge.

Lösch, A. (1954) *The Economics of Location: A pioneer book in the relations between economic goods and geography*, translated from the Second Revised (1944) Edition by William H. Woglom with the Assistance of Wolfgang F. Stolper, New Haven, CT: Yale University Press.

Lowder, S.K., Carisma, B. and Skoet, J. (2012) *Who Invests in Agriculture and How Much? An empirical review of the relative size of various investments in agriculture in low and middle income countries*, ESA Working Paper No. 12-09, Rome: FAO.

Marsden, T.K. (2012) 'Food systems under pressure: Regulatory instabilities and the challenge of sustainable development', in G. Spaargaren, P. Oosterveer and A. Loeber (eds) *Food Practices in Transition: Changing food consumption, retail and production in the age of reflexive modernity*, Routledge Series in Sustainability Transitions, New York: Routledge, pp. 291–312.

Mekonnen, M.M. and Hoekstra, A.Y. (2011) *National Water Food Print Accounts: The green, blue and grey water footprint of production and consumption. Vol. 1: Main Report*, Value of Water Research Report Series N.50, Delft: UNESCO Institute for Water Education.

Najam, A. (1999) 'World Business Council for Sustainable Development: The greening of business or a greenwash?', in H.O. Bergensen, G. Parmann and O.B. Thommessen (eds) *Yearbook of International Co-operation on Environment and Development 1999/2000*, London: Earthscan, pp. 65–75.

OECD (2012) *Agricultural Policy Monitoring and Evaluation 2012*, Paris: OECD Publishing.

Pimental, D. and Pimental, M. (2003) 'Sustainability of meat-based and plant-based diets and the environment', *The American Journal of Clinical Nutrition*, vol. 78, no. 3, pp. 660S–663S.

Popkin, B.M. (2001) 'The nutrition transition and obesity in the developing world', *The Journal of Nutrition*, vol. 131, no. 3, pp. 871S–873S.

Pretty, J., Sutherland, W. J., Ashby, J., Auburn, J., Baulcombe, D., Bell, M. and Toulmin, C. (2010) 'The top 100 questions of importance to the future of global agriculture', *International Journal of Agricultural Sustainability*, vol. 8, no. 4, pp. 219–236.

Soil Association (2008) *An Inconvenient Truth about Food: Neither secure nor resilient*, Bristol: Soil Association.

Soil Association (2009) *Food Futures: Strategies for resilient food and farming*, Bristol: Soil Association.

Soil Association (2012) *Feeding the Future: How organic farming can help feed the world*, Bristol: Soil Association.

Taleb, N.N. (2008) *The Black Swan: The impact of the highly improbable*, London: Penguin.

Tomlinson, I. (2011) 'Doubling food production to feed the 9 billion: A critical perspective on a key discourse of food security in the UK', *Journal of Rural Studies*, vol. 29, pp. 81–90.

UNDESA (2013) *World Population Prospects: The 2010 Revision and World Urbanization Prospects: The 2011 Revision*, New York: Population Division of the Department of Economic and Social Affairs of the United Nations Secretariat.

UN Global Compact (2012) *Scaling up Global Food Security and Sustainable Agriculture*, New York: United Nations Global Compact Office.

USDA (2010) *Local Food Systems: Concepts, impacts, and issues, economic research report no. (ERR-97)*, Washington, DC: United Nations Department of Agriculture.

WBCSD (2010) *Vision 2050*, Geneva: World Business Council for Sustainable Development.

Willer, H. and Kilcher, L. (eds) (2012) *The World of Organic Agriculture: Statistics and emerging trends 2012*, Bonn: Research Institute of Organic Agriculture (FiBL), Frick and International Federation of Organic Agriculture Movements (IFOAM).

Wilson, G.A. (2002) 'From productivism to post-productivism ... and back again? Exploring the (un) changed natural and mental landscapes of European agriculture', *Transactions of the Institute of British Geographers*, vol. 26, no. 1, pp. 77–102.

World Bank (2008) *World Development Report 2008: Agriculture for development*, Washington, DC: The World Bank.

World Bank (2010) *World Development Report 2010: Development and climate change*, Washington, DC: The World Bank.

World Bank (2012) *Food Price Watch. August 2012*, Washington DC: World Bank.

3 European food governance

The contrary influences of market liberalization and agricultural exceptionalism

Robert Lee

Introduction

Devised in the 1950s as a common market, the *raison d'être* of what is now the European Union was the free movement of goods. Given the significance of agri-food products in Europe, it was vital that trade in this area both within and outside of the Single European Market was liberalized. Yet, time and again, the food sector has been the location of disputes both between member states of Europe and between Europe and its external trading partners. There are a number of reasons for this. To begin with all member states had developed systems of food safety regulation often in advance of EU market entry. This is unsurprising as the social costs of unwholesome and particularly unsafe or unhygienic food are considerable. However, the market interventions of the regulatory state could prove highly protectionist in both design and implementation. Some of the great battlegrounds of European law of free movement of goods have been disputes concerning foodstuffs – think of French cassis (*Rewe-Zentral A.G.* v. *Bundesmonopolverwaltung fur Brannt-wein* ('Cassis de Dijon')) (Case 120/78 1979), French poultry (*Commission* v. *UK* (Case 40/82) 1982) or Danish beer (*Commission* v. *Denmark* (Case 302/86) 1988). Moreover, these national systems of food regulation were highly fragmented, not merely lacking uniformity in their construction but even serving differing objectives in balancing cultural interests in food against objectives of public health.

This chapter reviews the current framework for the governance of food within Europe but suggests that although food regulation has been centralized at a European level, it remains highly disjointed in its scope. The chapter begins by examining the division between agriculture and food and suggests that Europe has developed separate realms in which it addresses food producers and food consumers. It will attempt to show that this is unhelpful when dealing with a sector that, in globalized markets and under increasing pressure on resources to serve the food system, demands an integrated approach. This lack of integration is best highlighted by the inherent tension between a tradition of agricultural exceptionalism in food production and a commitment to market liberalization in food products in which the desire to promote the free movement of foodstuffs dominates all other considerations of food production and consumption. After reviewing issues of food consumption and consumer protection, this chapter explores the centralization of food governance in Europe based on regulatory science, and goes on to suggest that this framework above all serves the cause of market liberalization of food at the expense of other pressing concerns. These concerns are considered in the section entitled 'Towards the integrated governance of the agri-food system?' on p. 76 and in the conclusion. We begin, however, at the point of production, considering the exceptional position of agriculture in an otherwise liberalized European market.

The tradition of agricultural exceptionalism

The EU has long pursued highly discrete policies in relation to food production; whereas, as explained below, Europe-wide concerns with questions of food consumption are of more recent origin. From the very outset of the common market, an agricultural policy was thought vital to ensure food security in post-war Europe. The economic factors behind a common agricultural policy (CAP) stem from claims regarding the extraordinary features of agri-food both in terms of demand and supply (Nedergaard, 2006a). On the demand side, these include a recognition of the low price elasticity commonly associated with goods thought of as necessities. Increases in household income may not lead to significant increases in the proportion spent on food, since, even at a lower income level, spending on food is a high priority within the household budget. Supply side considerations begin with a perception of markets for agri-foods as unstable, since failures of harvests could significantly reduce supply. Farmers therefore work in a risk market for products that may be highly perishable. This limits the scope for flexibility in reacting to market changes over the growing and harvesting seasons for crops. It is this immobility in the face of price shifts that motivates attempts to bring stability and self-sufficiency to agri-food markets. In Europe in the light of post-war shortage, the justification for intervention was to overcome these perceived events of market failure, though we will later question whether the CAP solution is not equally a case of regulatory failure. Grace Skogstad has suggested that, in Europe, there is a recognition of the special interests of farm producers which grows out of a concern with national interest in food production to promote agricultural exceptionalism (Skogstad, 1998). Yet, ironically as food security is once again becoming a point of perturbation in European nation states, the protection of such special interests is a matter of global contestation.

In the UK, this relationship between state and farmer was explored in a book of that title that represented one of the earliest works on corporatism. Self and Storing (1962) explored the partnership of the state and the farmer, under the mediation of the National Farmers' Union pursuing mechanisms enshrined in the 1947 Agriculture Act. Under this model there was a corporatist bargain in which 'the Government has provided agriculture with a high degree of support and protection, and agriculture has accepted in return an unusual amount of supervision and regulation' (Self and Storing, 1962, p. 15). The exceptionalism could take many forms. One good example would be the regulation of agricultural workers' remuneration. At the time of writing (19 December 2012), the government has just announced the abolition of the long established Agricultural Wages Board for England and Wales together with related regional Committees in England by means of an amendment to the Enterprise and Regulatory Reform Bill. The amendment will also make changes to the National Minimum Wage Act to ensure that agricultural workers are protected by the National Minimum Wage once the agricultural minimum wage regime is abolished. The long-standing exclusion of agricultural waste from mainstream waste regulation might be another example; only since 2006 has agricultural waste been subject to the same controls as other sectors and, even now, it is mainly dealt with by way of waste exemptions. While much of the earlier corporatism within agricultural exceptionalism has died, the remains of the policy are still alive and kicking in places. By far the most prominent and most contentious heritage of agricultural exceptionalism is the area of subsidy for agricultural production – in Europe through the common agricultural policy.

Interestingly, although intervention in the form of a common agricultural policy might be seen as largely protecting producers from the vicissitudes of an uncertain market, the long-term consequence of the CAP appears to have been to contribute to a reduction of the number of European farmers not least by its implicit adoption of agricultural intensification (Robson, 1997). Notwithstanding the expansion of the European Community during this time, the forty years following 1957 and the Treaty of Rome saw a reduction in the number of European farmers of almost ten million persons. One reason why this might be so is that production subsidy also significantly supports non-farming actors in the food chain. For example the OECD has estimated that input suppliers have benefitted to a greater degree than producers and it is a common complaint in the UK that profits from farm produce shift from farmers to retailers. In the USA, two thirds of agricultural subsidy is routed to the top 10 per cent of producers. In Europe it is estimated that 80 per cent of subsidy goes to 20 per cent of recipients (Donald *et al.*, 2002). Note that these detrimental impacts of CAP are typical of a structure in which the stated objectives are often in opposition to each other. For example, the CAP looks to deliver value to consumers while maintaining farm incomes or to protect and stabilize domestic agri-food markets, while actually generating surplus, thereby increasing market volatility (USDA, 2001).

In any case, there must be doubts as to the extent to which the CAP can continue to offer protection from market forces. Even in the European Union, enlargement has meant increased competition through new market entrants (Swinnen, 2002), some of them largely agrarian economies such as Romania or Poland with much lower labour costs. Technology also has helped increase productive supply (Rosegrant *et al.*, 2008) across world markets as a whole, increasing the competitive forces on European farmers. This is combined with more recent uncertainties generated by climate change and, in certain areas of Europe, water pressures. Meanwhile environmental controls to protect water (such as the Nitrates Directive, 1991) or biodiversity (such as the Habitats Directive, 1992) may also affect yields. But if these strains indicate agriculture as a sector in relative decline, then it becomes even harder still to abandon policies of agricultural exceptionalism at this time (as the current round of 'reform' is discovering).

Nonetheless, immediately following the introduction of the CAP, Europe moved quickly from importer to exporter through to the generator of significant agricultural surplus by the time of the Uruguay Round of negotiations in the mid 1980s. The CAP had promoted a preference for EC production and had adopted trade measures as an overt mechanism to stabilize commodity prices in the European market. It was precisely this type of protectionist agricultural exceptionalism that the Uruguay Round sought to tackle. It addressed the exclusion of the agricultural sector from the General Agreement on Tariffs and Trade, the conclusion of which, in 1948, came at a time of concerns with national food security that drove a protectionist stance. Piecemeal negotiations on tariffs for particular agri-foods such as soya beans took place over time but in general the sector was accorded 'special treatment' gaining exemptions from rules on quantitative import restrictions and permitting agricultural subsidy, even for export (subject to a rule of 'equitable market share' which proved difficult to define or observe). The EU along with other developed countries took full advantage of GATT dispensations, often to deal with surplus capacity on home markets. Since the EU acted as buyer of last resort of agri-food surplus, it was in its interest to offer price support for exports so that, rather than increased supply of produce leading to a fall in prices, higher yields actually meant higher incomes in general. The total amount of export subsidies spent by all WTO members in 1998 was $5.4 billion of which the EU accounted for $4.95 billion (Panagariya, 2005). Export subsidy generated growing friction between the developed and developing worlds with 60 per cent of all trade disputes submitted to the GATT dispute settlement process between 1980 and 1990 relating to agriculture (Healy *et al*., 1998).

While agricultural exceptionalism grew with the European project for a significant period of time, as world markets became increasingly liberalized, it came to have wider international consequences. Although the Uruguay Round was successful in subjecting agriculture to GATT regulation, arguably tensions between the EU and USA were a more important factor in realizing this than the effect on the developing world of the protectionist policies of industrialized countries (Bureau *et al*., 2000). Nonetheless, there was for a time some recognition that the elimination of subsidy and tariffs would produce more effective world food markets and restore prices to a more realistic level, helping to end unstable and unsustainable practices engendered by inefficient levels of production on the back of subsidy. Such practices often denied the possibility of market access to developing nations that ought in any free market to have profited from Ricardo's law of comparative advantage (Ricardo, 1817; Sraffa, 1953).

Having taken seven years to reach any form of agreement, however, the achievement was limited. The Agreement contains generally applicable rules binding on WTO Members in the three areas of market access, domestic support and export subsidies. The Preamble of the Agreement seems to accept that this represents no more than a beginning since it sets as a longer term objective the establishment of a fair and market-orientated agricultural trading system with the aim of substantial progressive reductions in agricultural support and protection to correct restrictions and distortions in world agricultural markets. Article 20 of the Agreement also set 'the long term objective of substantial progressive reductions in support and protection, resulting in fundamental reform' with agreement to continue negotiations. These negotiations, through the Doha Round, were due to have concluded in 2010, but appear to have ended in stalemate (Cho, 2010). Although a development Round, the failure resulted in part from impasse between the United States and the EU, with the former unwilling to reduce agricultural subsidy without movement by the latter on lowering farm tariffs (Kennedy, 2008). These types of commitment to agricultural exceptionalism by the developed nations led the developing countries to believe that their own case for protection, to guard rural livelihoods and ensure affordable supplies of staple foodstuffs, was all the stronger and ultimately led to deadlock (Cho, 2010).

In Europe, following the end of the Uruguay Round in 1994, the reform of the CAP began. This was not solely, or perhaps even largely, driven by the Agriculture Agreement's commitment to liberalize the agricultural sector. It can best be understood in the light of the history of the CAP, which was no longer necessary, by this stage, to stimulate production. By the 1970s, increased productivity had moved the sector well into surplus production. Tackling surplus did not prove easy without reliance on export subsidy. This is well illustrated by the example of milk quotas, the introduction of which represented one of the first post-productivist attempts at CAP reform. Introduced in 1984 milk quotas were intended to last for a five-year period but were in fact extended until 1992. Each member state was given a quota which was then allocated among national producers, with a tax on surplus production. Yet with the minimum EU price set at a level higher than surrounding market prices, surplus continued requiring export refunds by way of grant to deal with these. The purchase or subsidy of surplus production naturally led to an increased CAP budget, which saw a 20bn ECU increase in the ten years between 1981 and 1991 (Scottish Government, 2003).

In addition to any international pressures post-Uruguay Round, then, there were budgetary constraints demanding reduced agricultural support; and a European Community study has suggested that net exporters of agricultural products are likely to offer lower levels of subsidy than net importers (European Commission Directorate General for Economic and

Financial Affairs, 1994). The MacSharry reforms of CAP in the mid 1990s shifted the emphasis from support prices to direct forms of payment (Grant, 1995; Kay, 1998; Lenschow, 1999). This broadly reduced the market price for certain commodities (e.g. cereals and beef) within the EU in an attempt to move closer to world pricing of these goods. The direct payments sought to compensate for the price reduction while rewarding other activity such as environmentally friendly stewardship of the land. Though these payments were supposed to be interim measures to soften the blow of price reduction, as world commodity prices began to fall during the 1990s, the direct payments continued, as did increases in the CAP budget. The argument presented in favour of the CAP is that it is helping to create a level playing field but based on higher standards of production rather than a lowest common denominator approach, with CAP payments offered as compensation for the higher standard required (foodpolitics.eu). However, in the course of this argument it is also asserted that in spite of an integrated market, member state domestic agricultural production must be preserved. The description of 'managed open markets' (foodpolitics.eu) is convenient but unconvincing.

Indeed the Agenda 2000 reforms of CAP, while attempting to again produce a better alignment with world prices by reducing guaranteed EU prices, overtly depended on the maintenance of direct payments to be administered by member states which had some freedom under subsidiarity principles to realize particular objectives for domestic agriculture. Broadly, however, enlargement meant greater use of structural and cohesion funds so that, conveniently, subsidy could be cloaked by support for cohesion and market integration. In the agricultural sector the return demanded for farm payments, in a move to a multi-functionality model (Skogstad, 2008), was environmental stewardship and landscape management. Swinbank and Daugbjerg (2006) have argued that decoupling the direct farm payments from production presented itself as a potential response to pressures on the CAP through the Doha Round. However, because up to 25 per cent of the single farm payment could be tied still to production at the election of the member state, the reform led to considerable fragmentation of the nature of CAP payments across the European market. Arguably, however, shifting payments away from production further divorces farming from food and renders an integrated agri-food policy in Europe a more remote prospect. This is because, rather than generating an integrated food policy, linking farm payments to the pursuit of other imperatives increases fragmentation and emphasizes the tension so that, while seemingly responding to global pressures to open up markets, the European Union clings on to exceptionalism, albeit in an adjusted form.

The greening of the CAP has become, according to Daugbjerg (2003), a policy end in its own right, divorced from pressures to liberalize markets as the pressures on the EU have abated while the Doha Round has stalled. The proposals for a post 2013 CAP further tie farm support to compliance with particular environmental goals. However, this is at the expense to some degree of the decoupling of subsidy and production since the environmental objectives are heavily tied to land use. Moreover, this rather entrenches agricultural exceptionalism, since it suggests that the positive externalities generated by rural land use management (such as drawing visitors to attractive rural landscapes (Cooper *et al.*, 2010) justifies market intervention and makes the case for subsidy. At the same time, however, we see in other policy areas, such as agricultural biotechnology, a desire to open up European markets, indicating that tensions between agricultural exceptionalism and market liberalization rather than being resolved are actually re-intensifying. This suggests that, while there is everything to play for still in the reform of the CAP, it is far too early to herald the abandonment of the tradition of agricultural exceptionalism, not least because of the emergence of a new round of food security concerns (Lee and Marsden, 2011).

Food and market liberalization: mobilizing foodstuffs under agricultural exceptionalism

Arguably, interests in food policy are no less narrowly drawn than those in agriculture and even then, according to Millstone and van Zwanenberg (2002, p. 595), 'consumer and health interests have routinely been subordinated to the objectives of furthering the commercial interests of farming and the food industry'. In relation to trade in foodstuffs, there is no hint of agricultural exceptionalism but rather a series of measures designed to centralize and liberalize any controls on the marketing of food within Europe. Such food policy as there has been has often been directed at allowing the free flow of foods, initially through a slow and unsuccessful attempt at harmonization of the composition of foods followed by mutual recognition of regulatory decisions within member states to allow foods onto the market (Bergeaud-Blacker and Ferretti, 2006). Alemanno (2006) has argued that, in this process, concern for free movement of goods overrode considerations of food safety. When consumer policy gradually developed in the EU, food became a concern of a wider policy of consumer protection rather than as part of an orchestrated agri-food policy. Over time, not only is the latter policy area highly fragmented but it becomes subject to changing and competing policy hierarchies in which agricultural exceptionalism is initially at the pinnacle before being subsumed by the market imperatives pursued by trade and consumer policies.

We have argued elsewhere (Marsden *et al.*, 2010) that a series of food crises, not least that of BSE, caused the European Union to reflect upon this fragmentation in policy and the low priority, to that point, given to questions of pan-European food safety (Vos, 2000). The initial signs of this included: the coordination of matters of food safety through meetings of commissioners discharging the various functions touching on food safety including agricultural policy, consumer protection, public health, internal market, etc.; and the centralization of responsibility under the consumer protection service (DG XXIV). Given the difficulties of political and policy management of a high degree of scientific uncertainty at the time of the BSE crisis (Jacob and Hellström, 2000), it is hardly surprising that considerable attention was paid to the linkage between what was seen as scientific risk assessment and institutional risk management. The risk assessment task was placed in the hands of a new European Food Safety Authority (EFSA) and was to be science-led, though the work of EFSA would not extend to questions of risk management, an issue to which we will return later.

The creation of EFSA and the passage of a new EU Food Safety Law implied a wresting of control from national agencies in member states and the significant centralization at an institutional level within the EU of the primary regulatory responsibility for the safety of food products. In this process the European Parliament, which began to place itself as the consumer champion, played an important role. Having produced a devastating report on the competence of the Commission in its management of food safety (European Parliament, 1997) and having threatened censure and even a vote of no-confidence as a result, and with a pressing need to restore the confidence of the European consumer in food, the European Parliament became a powerful player in ensuring reform of EU food safety regulation (Knowles *et al.*, 2007).

It helped, perhaps, that the food crises, such as BSE in the UK (van Zwanenberg and Millstone, 2003), the poisoning of Belgian poultry (Bernard *et al.*, 2002) and swine fever in the Netherlands (Elbers *et al.*, 1999), could be seen as nationally based. Even though day-to-day biosecurity and food safety controls would remain at member state level, this allowed an impetus in favour of an integrated European-wide system. Moreover, if the system was to be driven by 'sound science', then the argument was that a centralized agency could have access to the very best scientific thinking. However, the role of EFSA is confined to risk assessment with issues of risk management falling to the Commission. From the outset this was presented as an attempt to ring-fence an objective scientific analysis provided by EFSA on which risk managers in the Commission could then act (Marsden *et al.*, 2010). This separation is considered further below, but, in truth, it was driven to a large degree by pragmatic legal considerations because the transference of risk management powers to EFSA would have involved considerable rewriting of the institutional framework of the EU even to the point of Treaty change (European Commission, 2000).

The centralization of responsibility under DG Sanco following the food crises of the late twentieth century gave rise to a new consumerist turn in food policy and to some degree pushed questions of food production and its related exceptionalism even further into the background. It served to legitimate the management of food risks even though responsibility for risk management within Europe itself did not really change. It asserted the significance of science-based decision making, but this assumes that there is no ambiguity or contingency in how the science is to be interpreted in the process of risk management – an issue pursued in the section entitled 'Of science and separation' on p. 73. Not only is risk assessment divorced from risk management but risk communication lies still within member states and not merely with the competent authorities but also with private actors in the food chain. Thus, there is a centralized system but with fragmented responsibilities.

During the process of reform, the White Paper emphasized the 'essential role of the internal market' in the provision of quality foodstuffs as well as the significance of EFSA in assuring the economic aims of the European internal food market (European Commission, 2000). Moreover, in spite of the lead taken by the consumer protection service, DG Trade retained a strong interest in the risk assessment work of EFSA, if only because of their role in ensuring compliance with WTO rules and in particular perhaps the Sanitary and Phytosanitary (SPS) Agreement. Since biosecurity and other potential restrictions on trade in food must be based in science either by adopting accepted internationally recognized standards or following scientific risk assessment (Howse, 2000), the emphasis on a science-led approach in EFSA is hardly accidental. It is worth bearing in mind that coinciding with the period over which reform took place, there was a member-state imposed moratorium on approvals for GM products which was to trigger a WTO dispute and a dispute panel hearing (Winickoff *et al.*, 2005; *Biotech Products*, 2006). Even though a total of eighteen genetically modified organisms (GMOs) had been authorized for commercial release by October 1998, for the next six years the system of approvals ground to a halt after a number of member states threw their support behind calls from France and Greece for a moratorium (even though this was never formally approved).

In addition to this de facto cessation of approvals, a number of member states sought to resist the introduction of approved GMOs into their territory. They did so by invoking a safeguard provision in Directive 90/220/EEC (Deliberate Release of GMOs Directive, 1990) which allowed a state authority to restrict or prohibit the circulation of a GMO, approved by an authority in another member state, on grounds of an unacceptable risk to health or the environment. In none of the occasions on which this was done did the Scientific Committees endorse the risk assessment case put forward by member state agencies (Lee, 2005). In short, member states employed the safeguard clause as a means of asserting their unease at the commercial release of GMOs, often invoking the precautionary principle (Levidow, 2001), while the EU attempted to mobilize the Scientific Committee on Food as a bulwark to member state objections. Since such objections were borne out of a diverse range of concerns including protection of the environment and of local production and food cultures, it is unsurprising that cases for prohibition of GMOs based on the safeguard provision, which is premised on harm to public health, failed before the Scientific Committee.

Against this background, there is a different complexion to the creation of EFSA and the centralization of issues of risk assessment than if the reform is perceived only as a response to recurrent nationally based food crises. Instead, EFSA may be seen as a response to the protracted quarrel within Europe over GM crops and to internal as well as external perturbations in the legitimacy of agri-food policy. It may be that the Commission hoped that the sound science of EFSA would dilute member state resistance. This has not yet proved to be the case and the Commission have been forced to change tack by allowing, in the name of subsidiarity, state autonomy in decision making about their cultivation of GMOs. A proposed Article 26b to be inserted into Directive 2001/18/EC on the deliberate release into the environment of genetically modified organisms (Deliberate Release of GMOs Directive, 2001) would allow member states the choice of whether or not to cultivate GM crops. However, approvals of which crops might be cultivated would remain a matter of risk management for the Commission based upon scientific risk assessments conducted by EFSA. Ultimately, therefore, the construction of a centralized authority may have proved crucial in ending the impasse on the intractable problem of GM cultivation, allowing GM foods to circulate on the European market subject to controls on labelling and tracing (Carson and Lee, 2005).

From all of this it is possible to draw two sets of observations: one relating to the nature of the regulation and the other relating to its underpinning purposes. These are now examined in turn.

Regulation and disputation

Arising in part out of various food crises, including the conflict over GM, food has become characterized as an area of contested governance which Ansell and Vogel have described as 'associated with a pervasive sense of distrust that challenges the legitimacy of existing institutional arrangements' (Ansell and Vogel, 2006). As these authors point out, part of this contestation arises from the multilevel governance to which the reforms gave rise. As stated above, food safety is traditionally an area of national sovereignty so that one might expect some degree of resentment at the centralization of ultimate regulatory authority governing the circulation of food and, at least in some member states, that argument becomes wrapped up in concerns about the wider European agenda of what Ansell and Vogel depict as 'regulatory federalism'. This is a process that has moved from attempts to harmonize national laws, through to systems of mutual recognition based on harmonization of regulatory essentials, through to an increasing emphasis (in medicines, chemicals, intellectual property, etc.) on centralized European authorities (Bergeaud-Blacker and Ferretti, 2006). There may be an inevitable shift to multilevel governance in a system that is increasingly dominated by an international regime on free trade but this does not mean that the modification of legal authority is any easier to accept, and the world trade disputes involving the EU in the area of food are a clear indication of this.

As is the case with CAP reform, and as explained earlier in relation to the failure of the Doha Round, the story is often one of Europe in the form of the Commission, in particular, looking to adjust to international legal requirements without the guarantee of support, not merely among member states but also within the European Parliament (Nedergaard, 2006b, 2008). Examples of this might include the WTO beef hormones dispute precipitated by a Parliament-led ban that might be described as precautionary, but that without the requisite risk assessment led to a finding of infringement of world trade rules. A more recent example might be the heavy level of criticism directed by Parliament to the Commission for what was seen as a too facilitative and insufficiently precautionary stance on the use of nanomaterials in various regulatory settings including food. Such episodes help us to understand the preference in the creation and re-enforcement of EFSA for a science-led approach, but at the expense of placing science itself at the centre of distrust in this much contested governance model.

Regulation and liberalization

This contestation may be exacerbated by the mixed motives of the regulatory endeavour since food governance is not solely concerned with questions of food safety but is at least as involved with questions of trade and the free movement of goods in the single market. The GM regulation on the labelling and tracing of GM foods can be seen as facilitative of this trade (Stokes, 2012) in the same way as the introduction of autonomy over decision making on GM cultivation will allow the introduction of GM-based biotechnology into at least some parts of the European market (Wickson and Wynne, 2012). Moreover, regulatory devices such as tracing and labelling depend entirely upon the retail sector policing its supply lines, an activity in which it had commonly engaged even prior to this demand (Marsden *et al.*, 2000), if only for quality assurance purposes. Such dependence embeds the de facto private interest model of food governance, which arose originally out of failures in a public interest model (Marsden *et al.*, 2000) – *de jure* – as part of a new structure that recognizes the capacity of international retail supply chains to deliver regulatory objectives. The emergence of a powerful retailer-led model, which became the harbinger for the increased free flow of goods, further marginalized agricultural exeptionalism. Indeed, the production sector was expected to bear the brunt of both food regulation and value abstraction in this new phase so that, to the extent that these obligations are passed upstream through the supply chain, the burden falls upon producers in a manner that militates against any agricultural exceptionalism in the attempt to facilitate the free movement of these products in the market.

Indeed, in shaping the new legislative and institutional structures for a centralized system of European food governance at the heart of the enterprise lay a commitment to ensure the effective workings of such retailer-led supply chains within the internal market. While seeking to reassure the European consumer, the centralization can be seen as a ring-fencing of competitive territory for food markets and food trade. In our earlier work on food regulation, we make the point through empirical study that the consumer voice in food was largely constructed and financially supported by the EU institutions

(Marsden *et al.*, 2010). However, its recent emergence and its dependency has meant that, however strongly it presses its concerns for safe food, animal welfare and the like, it is ranked against a powerful private, commercial interest in safeguarding the effective workings of the internal market.

Meanwhile, the regulatory framework in Europe is now complex, hybrid (public/private) and relying on multilevels of governance. Many of these features emerge as a response to the imperative of promoting 'free' markets both inside and outside Europe, notwithstanding the uncomfortable fact that it is the underpinning anti-market tradition of agricultural exceptionalism that has driven the capacity to trade in agri-food products. Although there is a concern with agriculture, expressed through CAP at the outset of the common market, with the coming of the single market we see the emergence of food governance systems that are deeply embedded and integrated in a manner that was never quite true for agriculture simply because its status was exceptional. Moreover, the commercial and political significance of food as a policy arena has advanced as agricultural corporatism has declined. The latter has been to some extent replaced by the accommodation of new players such as retailers and consumers within a hybrid governance system that is complex in its networks and relationships but structured in its standardized and scientific approach.

This emphasis on market liberalization in relation to food as a product contrasts sharply with the agricultural exceptionalist approach to its mode of production. So different are these approaches that a unified approach to agri-food has proven to be highly elusive in Europe. Moreover, a commitment to science-based communication, between EFSA and, through institutional risk management, the European consumer, suggests a primary goal of stabilizing food markets by growing confidence in food safety while side-lining other concerns of food security and food production. Before examining this lack of integration, it is to this science-based approach that we now turn, since there are elements here that will aid our call for a more integrated approach to agri-food governance in Europe.

Of science and separation

While the food governance structures have taken shape and settled within the twenty-first century into a seemingly rigid bifurcation of risk assessment and risk management responsibilities, the nature of that risk is a matter of great complexity. The 'risk' may be better portrayed as a series of dynamic challenges, the dominant features of which are indeterminacy and contingency. Indeed, it is the need to grapple with such uncertainties, as exhibited by food crises, that has instigated the reconfigurations of power relationships within the food governance systems. This has led not merely to greater centralization of regulation within the EU but to the development of much more elaborate governance systems that extend across a wider range of jurisdictions and interests.

Central, however, is the role of EFSA, a body that grew out of scientific committees, and the idea of a repository of scientific expertise encompassing excellent and independent scientists free from non-scientific pressures and able to make decisions based on sound science (Alemanno, 2006). Through isolating questions of risk assessment so that the work of EFSA could then be taken up by risk managers, the danger of this split is that it merely entrenches the former model by privileging science as an autonomous domain able to operate apart from any wider social context attaching to food production and its safety. Ironically, the assertion of the neutrality of science decision making added considerably to the problems of dealing with BSE since, in fact, the science was employed to legitimate policy options to which the government continued to adhere in the belief that these would prove to be the most palatable politically (Millstone and van Zwanenberg, 2001). Departing from a model of an integrated regulatory authority along the lines of the UK Food Standards Agency, the danger of the split approach is that it becomes very difficult for other agencies, either at a central level within the EU or nationally within member states to depart from what is asserted as an independent, expert and authoritative determination. It also suggests a one size fits all approach in which locally based considerations concerning food safety issues are marginalized and in which there is little room for a precautionary stance in opposition to the EFSA assessment of risk.

It might be thought, against this view, that there would still be room for precaution at the risk management stage and that this remains possible is clear from the case of *Pfizer* (2002) in the European Court of Justice, where the fact that assessments suggested risks at a tolerable level did not bar the adoption of a precautionary approach by the decision makers. In part this is because the Court itself will not 'substitute its assessment of scientific and technical facts for that of the legislature on which the Treaty has placed that task' (*Afton Chemical*, 2010, para 28). However, at a policy level, and particularly when involving national agencies, such precautionary approaches are unwelcome and unlikely. Wickson and Wynne (2012) point to an instance of this relating to the proposed Article 26b of Directive 2001/18/EC (discussed above) and the freedom of member states to choose not to cultivate GM crops. The new proposal would allow bans only on 'grounds other than those related to the assessment of the adverse effect on health and environment' (European Commission, 2010). As the authors comment: 'The implication is that the EFSA adequately assesses health and environmental risks. Yet, this was, and remains, precisely the main issue for those member states that refuse to accept the scientific adequacy of EFSA authorizations' (Wickson and Wynne, 2012).

The division of risk assessment and risk management between different actors claiming different expertise appears flawed if both stages of this process are examined. The neutrality of the scientific stage holds obvious appeal because 'if science is objective and if it is value-free, it can arbitrate between competing views of social policy options' (Brunk *et al.*, 1991). This convenient resort to what has been described as 'regulatory science' (Jasonoff, 1995) ignores certain obvious difficulties that arise out of the value-laden nature of the enterprise once science is placed at the heart of regulatory decision making. Among these difficulties is the laying aside of contingency in order to give a regulatory lead. The very characterization of risks, the setting of criteria and the construal of effects are simple examples of necessary choices that will influence risk assessment outcomes. The choice of these and of other possible frameworks needs to be acknowledged, as do other normative assumptions surrounding methods, data, baselines and the like. All of these factors derogate from the notion of a single, certain answer that is exhaustive in its assessment of risk. If this is so, not only is the notion of an authority capable of generating definitive answers that will drive regulatory solutions exposed as idealized, but the concept of charging the risk assessment task exclusively to an expert body also appears deficient. There may be a range of other participants capable of bringing to the risk assessment table bodies of knowledge (including lay knowledge) that might help frame the risk assessment task.

If, however, the isolation of other voices from scientific risk assessment is unfortunate, so too is the separation out of risk management from the risk assessment upon which it is meant to be based. Presumably the idea is that, at this stage, a wider range of factors can be brought to bear in making risk management decisions. These could include: public health or animal health issues; environmental considerations; and other regulatory questions ranging beyond food safety, not least the effects on trade inside and outside of the European market. It seems idle to think, however, that these contexts might not also have aided risk assessment and opened up wider categorization of risk, since determining what is 'at risk' is a crucial first step in any process of risk assessment. But more than this; the earlier decision on risk assessment will now shape the risk management. Yet, if one accepts that EFSA will not be producing, in the nature of the task, a single, universal and unambiguous exposé of the risk, then might it not be helpful to have the contours of risk articulated and explored also at the risk management stage?

All of this suggests three problems for the European food governance model. The first of these, if there cannot be a process that will isolate and uncover risk through scientific risk assessment on which risk managers can then unequivocally act, is how are food safety decisions to be reached? The truth is that there will be no simple answers about how to isolate questions of safety from other questions of how we operate the food system. For example, production methods including agricultural intensification, patterns of breeding, and the long range transportation of live animals might all increase biosecurity risk (see Enticott, Chapter 6, this volume). This being so, the entire process of risk governance is better seen as a political and participative endeavour rather than a form of science-led administrative action. Second, centralization of food safety in Europe is also open to question. If the risk assessment conducted by EFSA is not a once and for all proclamation of factors on which we should act, and if we cannot know the import of inevitable contingencies or knowledge shortfalls, then there could be many other appropriate forums for

European risk governance. In particular, local, regional or national agencies may be better placed to understand settings of risk choices. Finally, the notion of an approach that is science-led by an agenda set by EFSA may demote a wide range of other factors that ought to be considered in the development of an integrated food system that might then begin to bridge the division between agricultural exceptionalism and food market liberalization. It is to these factors that we now turn.

Towards the integrated governance of the agri-food system?

It is not difficult to find policy areas that are capable of clashing with that for agri-food and we now offer three illustrations. One immediate example might be that of *energy and biofuels*. Under the Biofuels Directive (2003) the EU set a goal of reaching a 5.75 per cent share of renewable energy in the transport sector by 2010. By this date first-generation biofuels, mainly from crops such as sugar, starch or vegetable oils, accounted for about 4.7 per cent of EU transport fuel consumption. Around this time, the Renewables Directive (2009) saw this target shift to a minimum 10 per cent in every member state by 2020. Yet by the time the Directive passed into law, food prices in the European market had begun to rise. In October 2012, the Commission announced that it would limit the contribution of crop-based biofuels towards the renewable energy target. The fear was that use of crops for biofuels was at the expense of food production (at a time of rising food prices) and was putting areas of forest and peatland (carbon sinks) under pressure created by incentives for crop production. This episode illustrates that finite resources are now, especially as the 'low carbon agenda' emerges, in competition to serve differing economic needs and that adverse synergetic effects of changes in one policy area might be generated with surprising rapidity.

A second example comes from a field in which the EU has been particularly active, namely *environmental protection*. This has become a point of tension between the EU and the UK. Announcing his plans for an in–out referendum on Europe, David Cameron (2013) said that 'we need to examine whether the balance is right in so many areas where the European Union has legislated including on the environment'. This reference to the burdens of EU environmental law follows George Osborne's Autumn Statement in 2011 attacking the protection of EU scheduled habitats as imposing ridiculous costs on British industry (Juniper, 2013). Notwithstanding that this was shown not to be so by the UK Department for Environment, Food and Rural Affairs, what irks the British Government is that so much effort has been put into environmental protection by the EU institutions that 80 per cent of UK environmental law is said to emanate from Europe. Yet, ironically, the driving force of EU environmental law historically has been the wish to create a level playing field of regulation for European industry to serve the cause of market liberalization to the advantage of business interests (see also Miele and Lever, Chapter 7, this volume). In the case of agriculture, however, Buller (2002,

p. 120) has argued that 'the farming sector would appear to escape from the universal application of obligatory command and control legislation that is applied in other sectors'. Buller (2002, p. 103) would see measures such as the inclusion of the most intensive agricultural units within integrated pollution control as tinkering at the margin and suggests that policy in the EU has been based on an assumption that 'what was good for farming would necessarily be good for the countryside (in the widest sense of that word)'. Yet Ruhl (2000, p. 274) has suggested that while farming is highly dependent on natural resources, it is not always mindful of its stewardship under the tolerance of agricultural exceptionalism, detailing its impacts as including: habitat degradation; soil erosion and salinization; water resource depletion and pollution; chemical releases and air pollution; and inadequate animal waste disposal. Although EU commitment to environmental protection is sufficiently strong as to engender resistance in member states such as the UK, it appears to grind to a halt under forces of agricultural exceptionalism.

A third and final illustration of a policy area with obvious links to food is that of *health*. A great deal of time is spent within both member states and the EU itself considering questions of nutrition and health. For example, in May 2007, the Commission published a White Paper to launch 'a coherent and comprehensive Community strategy' to address issues of diet and obesity (European Commission, 2007). The White Paper points to 'evidence that good practices of retailers in member states, promoting healthy products such as fruits and vegetables at cheap prices on a regular basis, have led to a positive impact on diets'. It advocates promoting children's consumption of fruit and vegetables by allowing surplus production to be distributed to educational institutions. Yet the surplus exists largely because of the CAP which imposes import tariffs on imports of fruit and vegetables while guaranteeing a minimum price for EU produce. This entails financing the withdrawal from the market of excess fruits and vegetables. This is the surplus to which the White Paper refers. It is of course far better that it should go to school children rather than be destroyed, but it exists quite simply because of a policy to maintain high prices by limiting the market circulation of fruit and vegetables. Veerman *et al*. (2006) have argued convincingly that pursuing such CAP policies on fruit and vegetable products has negative health impacts when compared with health outcomes for the Dutch population of the withdrawal of CAP support in this area. This is a striking example of agricultural exceptionalism functioning in a manner contrary to wider objectives of food policy, though ironically, as we saw at the beginning of the chapter, the basis of the case for agricultural exceptionalism has been all about food as a basic human need.

These sorts of examples have led to increasing calls in a macro policy context of climate change, a low carbon economy, resource depletion, and food/water security for a much more integrated approach to food governance issues that would embrace multiple policy goals to promote sustainability (Lee and Marsden, 2009; Schneider, 2010). Schneider, writing of the USA, has called for 'a return to the agrarianism that reconciles the self interest of farmers with the public good of society (as the) hallmark of a new food-based agriculture' (pp. 946–947), pointing to a mutuality of interest between producers and consumers in the production, assurance and availability of healthy food. At the time of writing, the media in the UK and Ireland is preoccupied by the discovery of horse meat in processed meat products. There are many dimensions to this story, including the complexity of supply chains (as the first detected horsemeat came to the UK via – at least – Ireland and Poland) and potential health impacts (since the horsemeat tested positive for phenylbutarone, an anti-inflammatory drug used to treat lameness that is banned from the human food chain in the EU). However, the episode also demonstrates that multifaceted cultural and ethical questions surround food production and consumption in a way that would be untrue for many products that might circulate freely across markets.

In our work on a food strategy for Wales (Welsh Government, 2010) a team from BRASS, including a number of the authors of chapters in this book, explored the intersections of food, agriculture and other policy areas including health, food culture and education, food security, environmental sustainability, tourism, transportation, and community and rural development. The strategy called for an integrated approach to food policy in Wales, which should reconnect many of these issues to build resilience in the Welsh food system to the advantage of the Welsh economy by increasing capacity to compete in food markets. This work followed earlier work on food security in the UK conducted within the think tank, Chatham House (Ambler-Edwards *et al*., 2009). That work in particular focused on resource inputs into the food system and the fragile reliance on that system on increasingly pressurized natural capital, arguing that food security is now a pressing concern even in the developed world economies. Yet market liberalization, and the assumption that food importation will always deliver solutions, and agricultural exceptionalism, and its artificial generation of surplus at huge cost to the European taxpayer, each in their own way derogates from any serious attempt to address questions of European food security.

Conclusion: food security and sustainability falling through the gaps

In this chapter we have sought to argue that there are competing and oppositional policies of agricultural exceptionalism and market liberalization. The inherent tensions that these engender make integrated food governance in Europe a much more intricate issue than it otherwise might be. As a result, the EU finds itself now having to confront both sustainability and food security in a much more serious way, and yet remains a long way from any sensible approach to a sustainable and integrated policy for agri-food. Although reforms to the process of food regulation in Europe are relatively recent, a product of this millennium, an opportunity was missed to consider a more holistic approach. With hindsight, concerns to promote market liberalization in the face of consumer-led crises, relating to both food safety and the acceptance of GM biotechnology, actually narrowed the focus of reform towards a science-led approach that drew and fixed attention away from a superfluity of concerns now facing the European agri-food system.

Such concerns appear to have been addressed in ways that are partial and disconnected. The tensions between subsidized agriculture and a free market system are glaring. The supposed economic fruits of market liberalization seem to have been insufficiently plentiful to allow the EU to abandon subsidy and end exceptionalism. Nor is it clear just what it is within the neoliberal market approach to food that is likely to address the twin problems within agri-food of sustainability and security. Indeed, as we see from the recent horsemeat saga, the food flows across highly liberalized markets seem to exacerbate rather than resolve such problems. Moreover, as the horsemeat episode shows, consumer concerns about food safety, quality and provenance are far from resolved, and interestingly, the centralized European governance of food has very little response to make because this is not an issue that is capable of resolution through regulatory science and because, practically, any regulation of the flouting of food regulation is likely to depend on national enforcement. In addition, such enforcement is public law based in a context in which the hybrid governance system and, in particular, the private ordering of the food supply chain can be shown to have failed.

All of this suggests that on the three food policy fronts of agri-production, free trade and consumer protection we are not much further forward. The EU hangs on to agricultural exceptionalist models that ought to be seen as oppositional to its central goals of a free market. Meanwhile, the free market imperative has driven the centralization of food regulation, yet we see that the regulatory model is confused and has limited capability to deliver consumer protection. Meanwhile, missing from this policy triumvirate is a pressing concern of the sustainability and security of agri-food systems that continually seems to fall between the gaps of a fragmented policy framework.

References

Afton Chemical Limited v. *Secretary of State for Transport* (2010) (Case C-343/09) 8 July 2010.

Alemanno, A. (2006) 'The evolution of European food regulation: Why the European Food Safety Authority is not a EU-style FDA?', in C. Ansell and D. Vogel (eds) *What's the Beef? The contested governance of European food safety*, Cambridge, MA: MIT Press, pp. 237–258.

Ambler-Edwards, S., Bailey, K., Kiff, A., Lang, T., Lee, R., Marsden, T., Simons, D. and Tibbs, H. (2009) *Food Futures: Rethinking UK strategy*, London: Chatham House.

Ansell, C. and Vogel, D. (2006) *What's the Beef? The contested governance of European food safety*, Cambridge, MA: MIT Press.

Bergeaud-Blacker, F. and Ferretti, M. (2006) 'More politics, stronger consumers? A new division of responsibility for food in the European Union', *Appetite*, vol. 47, no. 2, pp. 134–142.

Bernard, A., Broeckaert, F., De Poorter, G., De Cook, A., Hermans, C., Saegerman, C. and Houins, G. (2002) 'The Belgian PCB/dioxin incident: analysis of the food chain contamination and health risk evaluation', *Environmental Research*, vol. 88, pp. 1–18.

Biofuels Directive (2003) Directive 2003/30/EC of the European Parliament and of the Council of 8 May 2003 on the promotion of the use of biofuels or other renewable fuels for transport, OJEU/L123, 17 May 2003.

Biotech Products (2006) Reports of the Panel, *European Communities—Measures Affecting the Approval and Marketing of Biotech Products*, at 1, WT/DS291/R (U.S.), WT/DS292/R (Can.), WT/DS293/R (Arg.) (29 September).

Brunk, C.G., Haworth, L. and Lee, B. (1991) *Value Assumptions in Risk Assessment: A case study of the Alachlor controversy*, Waterloo, Canada: Wilfred Laurier University Press.

Buller, H. (2002) 'Integrating European Union environmental and agricultural policy', in A. Lenschow (ed.) *Environmental Policy Integration Greening Sectoral Policies in Europe*, London: Earthscan, pp. 103–126.

Bureau, J.C., Fulponi, L. and Salvatici, L. (2000) 'Comparing EU and US trade liberalisation under the Uruguay Round agreement on agriculture', *European Review of Agricultural Economics*, vol. 27, no. 3, pp. 259–280.

Cameron, D. (2013) Speech at Bloomberg, available at: www.bbc.co.uk/news/uk-politics-21158316, accessed 13 June 2013.

Carson, L. and Lee, R. (2005) 'Consumer sovereignty and the regulatory history of the European market for GM foods', *Environmental Law Review*, vol. 7, no. 3, pp. 173–189.

Cho, S. (2010) 'The demise of development in the Doha Round negotiations', *Texas International Law Journal*, vol. 31, pp. 573–601.

Commission v. *Denmark* (1988) (Case 302/86), ECR4607 ... 681–682.

Commission v. *UK* (1982) (Case 40/82), ECR2793 ... 307, 308.

Cooper, T., By, H. and Rayment, M. (2010) *Developing a More Comprehensive Rationale for EU Funding for the Environment*, Paper 3, London: Institute for European Environmental Policy.

Daugbjerg, C. (2003) 'Policy feedback and paradigm shift in EU agricultural policy: The effects of the MacSharry reform on future reform', *Journal of European Public Policy*, vol. 10, no. 3, pp. 421–437.

Deliberate Release of GMOs Directive (1990) Directive 90/220/EEC of 23 April 1990 on the deliberate release into the environment of genetically modified organisms.

Deliberate Release of GMOs Directive (2001) Directive 2001/18/EC of 12 March 2001 on the deliberate release into the environment of genetically modified organisms.

Donald, P.F., Pisano, G., Rayment, M.D. and Pain, D.J. (2002) 'The common agricultural policy, EU enlargement and the conservation of Europe's farmland birds', *Agriculture, Ecosystems and Environment*, vol. 89, no. 3, pp. 167–182.

Elbers, A.R.W., Stegeman, J.A., Moser, H., Ekker, H.M., Smak, J.A. and Pluimers, F.H. (1999) 'The classical swine fever epidemic 1997–1998 in the Netherlands: Descriptive epidemiology', *Preventive Veterinary Medicine*, vol. 42, no. 3–4, pp. 157–184.

European Commission (2000) *White Paper on Food Safety* COM (1999) 719 final.

European Commission (2007) *White paper on Nutrition, Overweight and Obesity Related Health Issues* COM (2007) 279 final.

European Commission (2010) *Proposal for a Regulation of the European Parliament and of the Council amending Directive 2001/18/EC as regards the possibility for the Member States to restrict or prohibit the cultivation of GMOs in their territory* COM (2010) 375 final.

European Commission Directorate General for Economic and Financial Affairs (1994) *EC Agricultural Policy for the 21st Century*, European Economy Reports and Studies 1994/4, Luxembourg: Office for the Official Publications of the European Community.

European Parliament (1997) Report of the Temporary Committee of Inquiry into BSE, set up by the Parliament in July 1996, on the alleged contraventions or maladministration in the implementation of community law in relation to BSE, without prejudice to the jurisdiction of the community and the national courts (February 7 1997, A4-0020/97/A, PE220.533/fin/A) – widely referred to under the name of the Committee Chair as the 'Medina Report'.

Grant, W. (1995) 'The limits of Common Agricultural Policy reform and the option for denationalization', *Journal of European Public Policy*, vol. 2, pp. 1–18.

Habitats Directive (1992) Council Directive 92/43/EEC of 21 May 1992 on the conservation of natural habitats and of wild fauna and flora.

Healy, S., Pearce, R. and Stockbridge, M. (1998) *The Implications of the Uruguay Round Agreement on Agriculture for Developing Countries*. Rome: Food and Agriculture Organization (FAO).

Howse, R. (2000) 'Democracy, science and free trade: Risk regulation on trial at the WTO', *University of Michigan Law Review*, vol. 29, pp. 2329–2357.

Jacob, M. and Hellström, T. (2000) 'Policy understanding of science, public trust and the BSE-CJD crisis', *Journal of Hazardous Materials*, vol. 78, no. 1–3, pp. 303–317.

Jasonoff, S. (1995) 'Procedural choices in regulatory science', *Technology in Society*, vol. 17, no. 3, pp. 279–293.

Juniper, T. (2013) 'David Cameron's EU speech is grave news for our environment', *The Guardian*, 25 January 2013.

Kay, A. (1998) *The Reform of the Common Agricultural Policy*, Oxford: CABI Publishing.

Kennedy, K. (2008) 'Doha Round negotiations on agricultural subsidies', *Denver Journal of International Law and Policy*, vol. 36, no. 3–4, pp. 335–348.

Knowles, T., Moody, R. and McEachern, M. (2007) 'European food scares and their impact on EU food policy', *British Food Journal*, vol. 109, no. 1, pp. 43–67.

Lee, R. (2005) 'GM resistant: Europe and the WTO panel dispute on biotech products', in J. Gunning and S. Holm (eds) *Ethics, Law, and Society*. Volume 1. Aldershot, UK, Burlington, VT: Ashgate, pp. 131–140.

Lee, R. and Marsden, T. (2009) 'The globalization and re-localization of material flows: Four phases of food regulation', *Journal of Law and Society*, vol. 36, no. 1, pp. 129–144.

Lee, R. and Marsden, T. (2011) 'Food futures: System transitions towards UK food security', *Journal of Human Rights and the Environment*, vol. 2, no. 2, pp. 201–216.

Lenschow, A. (1999) 'The greening of the EU', *Environment and Planning C*, vol. 17, no. 1, pp. 91–108.

Levidow, L. (2001) 'Precautionary uncertainty: Regulating GM crops in Europe', *Social Studies of Science*, vol. 31, no. 6, pp. 845–878.

Marsden, T., Flynn, A. and Harrison, M. (2000) *Consuming Interests: The social provision of foods*, London: UCL Press.

Marsden, T., Lee, R., Flynn, A. and Thankappan, S. (2010) *The New Regulation and Governance of Food*, Abingdon: Routledge.

Millstone, E. and Van Zwanenberg, P. (2001) 'The politics of expert advice: Lessons from the early history of the BSE saga', *Science and Public Policy*, vol. 28, no. 2, pp. 99–112.

Millstone, E. and Van Zwanenberg, P. (2002) 'The evolution of food safety policy-making institutions in the UK, EU and codex alimentarius', *Social Policy and Administration*, vol. 36, pp. 593–609.

Nedergaard, P. (2006a) 'Market failures and government failures: A theoretical model of the Common Agricultural Policy', *Public Choice*, vol. 127, no. 3–4, pp. 385–405.

Nedergaard, P. (2006b) 'The 2003 reform of the Common Agricultural Policy: Against all odds or rational explanations?', *Journal of European Integration*, vol. 28, no. 3, pp. 203–223.

Nedergaard, P. (2008) 'The reform of the Common Agricultural Policy: An advocacy coalition explanation', *Policy Studies*, vol. 29, no. 2, pp. 179–195.

Nitrates Directive (1991) Council Directive 91/676/EEC of 12 December 1991 concerning the protection of waters against pollution caused by nitrates from agricultural sources.

Panagariya, A. (2005) 'Liberalizing agriculture', *Foreign Affairs*, Special Edition, pp. 56–66.

Pfizer Animal Health v. *Council* (2002) T-13/99 ECR II3305.

Renewables Directive (2009) Directive 2009/28/EC of the European Parliament and of the Council of 23 April 2009 on the promotion of the use of energy from renewable sources.

Rewe-Zentral A.G. v. *Bundesmonopolverwaltung fur Brannt-wein* ('Cassis de Dijon') (1979) Case 120/78.

Ricardo, D. (1817) *On the Principles of Political Economy and Taxation*, John Murray, London, in P. Sraffa (ed.) (1953) *Ricardo: Works and Correspondence*, Cambridge: Cambridge University Press.

Robson, R.A. (1997) 'The evolution of the Common Agriculture Policy and the incorporation of environmental consideration', in D.J. Pain and M.W. Pienkowski (eds) *Farming and Birds in Europe: The Common Agricultural Policy and its implications for bird conservation*, London: Academic Press, pp. 43–78.

Rosegrant, M.W., Zhu, T., Msangi, S. and Sulser, T. (2008) 'Global scenarios for biofuels: Impacts and implications', *Review of Agricultural Economics*, vol. 30, pp. 495–505.

Ruhl, J.B. (2000) 'Farms, their environmental harms, and environmental law', *Ecology Law Quarterly*, vol. 27, no. 2, pp. 263–304.

Schneider, S.A. (2010) 'A reconsideration of agricultural law: A call for the law of food, farming, and sustainability', *William and Mary Environmental Law and Policy Review*, vol. 34, pp. 935–963.

Scottish Government (2003) *Reform of the Common Agricultural Policy: Discussion Paper* available at: www.scotland.gov.uk/Resource/Doc/1037/0003475.pdf, accessed 28 March 2013.

Self, P. and Storing, H. (1962) *The State and the Farmer*. London: Allen & Unwin.

Skogstad, G. (1998) 'Ideas, paradigms and institutions: Agricultural exceptionalism in the European Union and the United States', *Governance*, vol. 11, no. 4, pp. 463–490.

Skogstag, G. (2008) *Internationalization and Canadian Agriculture: Policy and governing paradigms*, Toronto: University of Toronto Press.

Sraffa, P. (ed.) (1953) *Ricardo: Works and correspondence*, Cambridge: Cambridge University Press.

Stokes, E. (2012) 'Nanotechnology and the products of inherited regulation', *Journal of Law and Society*, vol. 39, no. 1, pp. 9–112.

Swinbank, A. and Daugbjerg, C. (2006) 'The 2003 CAP reform: Accommodating WTO pressures', *Comparative European Politics*, vol. 4, no. 1, pp. 47–64.

Swinnen, J.F.M. (2002) 'Transition and integration in Europe: Implications for agricultural and food markets, policy and trade agreements', *The World Economy*, vol. 25, no. 4, pp. 481–501.

USDA (2001) *Agricultural Policy Reform in the WTO: The road ahead.* Agricultural Report, No. 802, United States Department of Agriculture.

Van Zwanenberg, P. and Millstone, E. (2003) 'BSE: A paradigm of policy failure', *Political Quarterly*, vol. 74, no. 1, pp. 27–37.

Veerman, J.L., Barendregt, J.J. and Mackenbach J.P. (2006) 'The European Common Agricultural Policy on fruits and vegetables: Exploring potential health gain from reform', *European Journal of Public Health*, vol. 16, no. 1, pp. 31–35.

Vos, E.I.L. (2000) 'EU food safety regulation in the aftermath of the BSE crisis', *Journal of Consumer Policy*, vol. 23, no. 3, pp. 227–255.

Welsh Assembly Government (2010) *Food for Wales, Food from Wales 2010:2020 – A Food Strategy for Wales*, Cardiff: Welsh Government.

Wickson, F. and Wynne, B. (2012) 'The anglerfish deception: The light of proposed reform in the regulation of GM crops hides underlying problems in EU science and governance', *EMBO Reports*, vol. 13, no. 2, pp. 100–105.

Winickoff, D., Jasanoff, S., Busch, L., Grove-White, R. and Wynne B. (2005) 'Adjudicating the GM food wars: Science, risk, and democracy in world trade law', *Yale Journal of International Law*, vol. 30, no. 1, pp. 81–124.

4 The public plate

Harnessing the power of purchase

Kevin Morgan and Adrian Morley

Introduction

Public procurement has moved from the margins to the mainstream in debates about sustainable food systems as analysts and advocates have belatedly realised that the power of purchase can help to fashion a healthier food system if it is deployed with skill and political will. Although all public sector organisations purchase food, schools, hospitals, care homes and prisons are the most prominent public food institutions, both in terms of the scale of their operations and their resonance with the general public. What schools, hospitals, care homes and prisons have in common is the fact that they nourish – or at least ought to nourish – the most vulnerable people in our societies, be they patients, pupils, pensioners or prisoners. In the UK alone, it is estimated that the public sector spends over £2 billion per annum on food and catering services, of which school meals are the largest category by value (Defra, 2012). Advocates of sustainable food procurement argue that this budget should be better deployed to encourage social, environmental and economic goals through the purchase of better quality food.

Procuring food from more sustainable sources typically means buying local, organic or fairly traded products from near and far, all of which can be part of a sustainable food narrative. This is an important point to make at the outset because sustainability, which embraces multiple values, needs to be framed in a capacious sense rather than being narrowly identified with a single attribute – be it the localism of the local food movement or the carbon-centric obsession of much of the green movement, an obsession that runs the risk of reducing sustainability to a glorified form of carbonism (Morgan, 2010).

Far from being a prosaic back office function, which has been the dominant perception in the past, public procurement is potentially one of the most powerful instruments that governments have at their disposal for effecting social, economic and environmental change (Meadowcroft, 2007; Morgan and Sonnino, 2008). The OECD estimates that the purchase of goods and services by public organisations accounts for, on average, around 15 per cent of GDP among its member countries and up to 70 per cent of GDP in developing countries (OECD, 2002). This presents us with a great paradox: while public procurement policy possesses enormous potential for effecting behavioural change in economy and society, policy makers have shown little or no interest in this policy until relatively recently.

This chapter draws on more than a decade of research by the authors on the prospects for what we have elsewhere called 'creative public procurement' (Morgan and Morley, 2002). As well as attempting to understand the theory and practice of public procurement, academic research has also sought to engage civil society advocates and policy stakeholders to promote creative/sustainable forms of procurement, particularly with respect to food. Indeed, one of the main impacts of academic research in this area has been to highlight 'good practice' initiatives, including how public food provisioning and sustainable procurement are regarded in different countries and regions of the world and among different stakeholder groups (Morgan and Sonnino, 2008). Understanding how other nations frame their school meals services, for instance, has helped practitioners and policy makers to re-imagine their own services because it exposes them to a wider array of possibilities for public food provisioning. A prime example would be the byzantine issue of EU public procurement regulations. Consistently held up as a major barrier to sustainable food procurement, these public procurement regulations can be interpreted in radically different ways and the great merit of comparative research has been to explain this diversity in terms of the dynamic interplay of culture and politics. That is to say, comparative research has shown that public bodies in the EU, all of which operate under the same regulatory framework, act in very different ways when it comes to interpreting EU regulations because they draw on different cultural values and enjoy different degrees of political support, and these factors help to account for the coexistence of leaders and laggards even in the same country (Morgan and Morley, 2002; Morgan and Sonnino, 2008).

The main aim of this chapter is to synthesise our thinking on sustainable food procurement in two distinct contexts: first as a professional practice that looms large in debates about public sector reform and, second, as a subject of academic debates about the changing nature of the public realm in an age of financial austerity. The following section outlines the development of public sector catering systems, taking the UK school meals system as an example. We then explore some of the challenges to the development of sustainable procurement theory, policy and practice. The final section distils the key factors (including the values) that need to be mobilised in a public procurement repertoire if the power of purchase is to be fully harnessed as part of a new sustainable food paradigm.

The evolution of public sector catering

School food is probably the most visible part of the public plate and the sector that resonates most strongly for politicians, policy makers and the general public. This resonance stems from the highly emotive nature of feeding vulnerable children, an activity that carries enormous implications for their health and well-being in the here and now as well as for their lifelong attitudes to food and diet. The school meals service is under growing pressure to nurture responsible attitudes to food choice because of the escalating costs of diet-related diseases such as obesity, a condition that already costs the UK National Health Service over £5 billion per annum. According to the WHO, there were more than 500 million obese people in the world in 2008, alongside a total of around 1.4 billion overweight adults (WHO, 2012). A more recent study put the total number of overweight children at 40 million (de Onis *et al.*, 2010). The OECD has warned that by 2020, 75 per cent of United States citizens could be clinically obese (OECD, 2010). The cost to society of dealing with the health implications of obesity is significant, especially in the US, which is estimated to spend over $147 billion annually on obesity-related medical care (Finkelstein *et al.*, 2009). In educational terms, it has been shown that school food can help to fashion a more congenial learning environment, yielding educational dividends even in poor areas such as the London borough of Greenwich (Belot and James, 2009).

Despite all the potential benefits of a successful school meal system, the sector has struggled to remain viable in many countries, especially in the hunger-blighted countries of the global South (Morgan *et al.*, 2007). At the heart of state sponsored school meal provision in the global North is the conflict between politically left-of-centre protagonists, who argue for the social benefits of publicly provided school food, and neoliberal detractors, who favour private choice rather than public provision. Nowhere was this conflict more apparent than in the British school meal system.

The roots of school meal provision in the UK can be traced back to the 1880s and the development of compulsory education. In particular, concern over undernourished children and their inability to learn effectively led to the Education (Provision of Meals) Act of 1906, which first gave local education providers the power to provide free school meals to pupils from poor backgrounds. Although nutritional standards were introduced in 1941, it was not until the latter days of the Second World War that the Education Act (1944) mandated the provision of school meals to all state-run primary and secondary schools in the country. Included in this legislation was the principle that school meals should not be profit making (i.e. the cost of provision to pupils should not exceed the cost of the service for school authorities) and that the standard of the school lunch should be sufficient for the main meal of the day. This was a welfare model of school food provisioning and it lasted until the 1970s, when a neoliberal model was introduced as part of the pro-market reforms of the Thatcherite era (Morgan and Sonnino, 2008).

The 1980 Education Act removed the mandate for schools to provide food to pupils beyond their free school meals obligation. It also ended both the nutritional standards and the requirement for schools to provide meals at a fixed price. The pressure on school food was further compounded by the introduction of compulsory competitive tendering (CCT) in 1988, a momentous moment in the history of public sector catering because it required local authorities to open services, including school meals, to the private sector and make service provision decisions based primarily on the lowest cost. The introduction of CCT, and its successor Best Value, had a significant impact on UK school meal provision. The high premium attached to low cost had two devastating effects on the school catering service: it led to the erosion of kitchen infrastructure and the deskilling of school cooks as the use of processed food products replaced the traditional skill set associated with cooking from scratch, precipitating a downward spiral of kitchen infrastructure, cooking skills and food that was described as 'processed muck' by one prominent school cook (Orrey, 2003).

As the costs of the neoliberal model became more evident – not least the financial costs of diet-related disease which are partly attributable to poor food choice at school – political and public opinion moved in favour of a new model of school food provisioning, a model that has been characterised as the ecological model because it seeks to integrate the principles of public health, social justice and ecological integrity (Morgan, 2008). Aided and abetted by a popular media campaign by celebrity chef Jamie Oliver, and by civil society advocacy groups such as Sustain and the Soil Association, school food policy in the UK began to address the 'whole school approach', which sought to align the *message* of the classroom (to eat well and be well) with the *menu* of the dining room (to ensure the quality of food underlined rather than undermined the message of the classroom). One of the earliest initiatives to champion the 'whole school approach' was the Scottish Government's *Hungry for Success* initiative in 2002 (Scottish Executive, 2002). As well as calling for nutrient-based standards for schools in Scotland, the initiative put the whole school approach at the centre of its new vision of school food reform, where the key argument was that school meals should be regarded more as a health service than a commercial one. This is an approach that has garnered increasing support from educationists, nutritionists, sustainable food advocates and the school meals sector itself, even if it remains limited in practice. Whole school approaches to school food recognise, and aim to harness, the interconnectivity between school food, child nutrition and educational attainment as well as wider public health, social justice and environmental sustainability issues. This new ecological approach to school meals can be viewed as being so distinct to the preceding CCT era that it should be regarded as nothing less than a 'school food revolution' (Morgan and Sonnino, 2008). Building on the vision of *Hungry for Success* in Scotland, school food reform in England has been pioneered by the Soil Association, which led the hugely influential coalition that produced the *Food for Life Partnership*, which is arguably the most ambitious school food reform ever undertaken in the UK, not least because it seeks to integrate the provision of good quality school food with education programmes for children and training programmes for cooks and food chain suppliers.

The ecological model of school food provisioning is struggling to survive in a mainstream public sector catering system in which low cost (if not lowest cost) continues to dominate the mindsets of catering and procurement managers (Morgan and Sonnino, 2008; Morgan, 2010). The fact that public sector food provision is routinely referred to as 'cost sector catering' within the industry highlights the ingrained emphasis on cost minimisation by purchasing officers and their political masters. Indeed, the new age of austerity triggered by the financial crisis of 2008 presents a formidable challenge because it threatens to reverse all the little victories that have been achieved in public sector catering over the previous decade (Morgan, 2010).

Two direct and interrelated effects flow from this low cost ethos: first, 'sustainable' food typically costs more and, second, the financial fragility of the system makes it susceptible to external shocks, such as falls in uptake (as happened after the 2005 Jamie Oliver television campaign) and cuts to subsidies. In addition, school caterers have to contend with rising food prices as well as escalating labour and energy costs. The only way the school meals service can become economically sustainable without recourse to external funding is either to increase the price charged to parents or to increase overall take-up rates for the service, thus reducing fixed overheads. It is estimated that average take-up levels of 55–60 per cent are needed before school meals services can break even and become self-financing. The average take-up in England was estimated at 46.3 per cent in primary schools and 39.8 per cent in secondary schools from 2011–2012 (School Food Trust, 2012). The goal of a self-sustaining service looks unlikely if local authorities reduce or withdraw their subsidies in the light of the funding crisis.

The progress of recent years is therefore in danger of being undone by a new cost-cutting drive. In a bid to counter these pressures, schools and Local Authorities are investigating new arrangements that bring either reduced costs or greater control. One such option is to scale up purchasing arrangements through buying consortia. In Wales, the Welsh Government is keen to promote organisational innovation along these lines so as to overcome the high costs associated with twenty-two separate local authorities working within their organisational silos. Scaling up does, however, present possible threats to sustainability more broadly, as purchasing through large contracts tends to favour large 'conventional' supply chain arrangements – the aggregation of demand tends to result in the aggregation of supply. While the implications of such aggregation are not entirely clear from a sustainability standpoint – because small and local suppliers do not necessarily equate with sustainable suppliers – it is clear that such systems ape the conventional supply chains and values that have contributed to the current sustainability crisis. Centralised systems also potentially challenge issues of social justice and the rights of citizens to define their own food sustainability.

The UK is, of course, not alone as a site of struggle for sustainability in institutional food systems, whether for cost or related reasons. An area of strong progress in recent years has been among US universities and colleges (Barlett, 2011) as well as schools, particularly with respect to local sourcing (Izumi *et al.*, 2010). A 2002 change in federal legislation mandated the US

Secretary of Agriculture to encourage the purchasing of locally grown food to the 'maximum extent practical and appropriate' by institutions within the federal school meals programme, while the 2008 Farm Bill supported this new direction by allowing schools to set geographical preferences when setting out purchasing policy (Izumi *et al.*, 2010). While these regulatory reforms are a step in the right direction, not least because the federal system begins to foster rather than frustrate school food reform, they are no substitute for robust community-based action at the level of states, cities and counties, the levels at which consumers and citizens can be actively involved in fashioning a more sustainable school food service.

The complexities of food procurement

As we have seen in the case of the British school meal system, public food provision is typically a highly contested arena that is exposed to competing pressures to be economical, nutritious, socially just, environmentally sustainable and organoleptically pleasurable. Striking a judicious balance between all these values, so as to realise the benefits in the here and now while respecting the needs of future generations, is the very essence of the quest for a sustainable public plate. However, some formidable barriers have to be overcome before sustainable procurement becomes a common practice in the public sector in the UK, as follows:

Box 4.1: Common barriers to sustainable procurement

- Cost: Perception of increased costs associated with sustainable procurement. Value for money is perceived to be inconsistent with paying a premium to achieve sustainability objectives.
- Knowledge: Lack of awareness of the need for and processes required to conduct procurement more sustainably.
- Awareness and information: Lack of information about the most sustainable option; lack of awareness of products; lack of monitoring of suppliers; perceptions of inferior quality.
- Risk: Risk-averse buyers prefer to purchase from suppliers with a good track record. Organisations fear criticism from the media and are therefore less keen to take innovative approaches.
- Legal issues: Uncertainty as to what can, and cannot be done, under existing rules (both UK and EC) on public procurement.
- Leadership: A lack of leadership – both organisational and political – leading to a lack of ownership and accountability at all levels.
- Inertia: Lack of appetite for change. Lack of personal or organisational incentives to drive change.

Source: National Audit Office (2005)

With the possible exception of cost, these barriers reflect a lack of professional skills to frame the procurement challenge in a creative fashion and the equally important lack of political confidence to transform existing practice. However, as countless good practice initiatives illustrate, these barriers are surmountable. Even the issue of cost, which looms very large in public food provisioning, can be addressed through creative purchasing policy and judicious menu planning as the City of Malmö has clearly demonstrated (Morgan, 2014).

As this chapter will illustrate, the mainstreaming of sustainable procurement depends on removing these barriers across the sector. One of the greatest barriers to this is the fact that, far from being the homogenous entity that its name seems to imply, the public sector is actually highly fragmented with an array of sub-sectors – such as health, education, care homes, local government, prisons, for example – and these sub-sectors have not been in the habit of collaborating to find joint solutions to common problems. In short, the fragmented governance structure of the public sector is one of the most intractable barriers to the dissemination of good procurement practice.

Take local government for example. In addition to procurement and catering managers, there is a wide array of positions and departments dealing with finance, sustainability, economic development and public health, each with its own priorities, budgets and reporting lines, not to mention the political dynamics of elected representatives and the local electorate. Furthermore, within the school environment, there is a whole series of additional constituencies that need to be factored into the school food reform equation, principally school governors, teaching staff, pupils, parents, as well as food suppliers and kitchen staff.

On the face of it the school food service appears to be a simple confection, but this prosaic image conceals a degree of complexity that belies its appearance. It is this complexity that has stymied the process of school food reform and rendered the research task of understanding the process more challenging. Indeed, the history of school food reform is littered with examples of initiatives that have been undone by two factors in particular: (i) reformers have consistently underestimated the level of complexity in the school food system and (ii) reformers in civil society organisations have found that key stakeholders in local government were either unable or unwilling to commit to genuine school food reform (see Morgan and Morley, 2006; Morgan and Sonnino, 2008; Morley *et al.*, 2012). Barlett (2011) describes the pioneering developments at Hendrix College, Arkansas, during the 1980s, to build quality into its catering system, including sourcing up to 30 per cent of food from local farmers. Unfortunately, a very common thing happened to this initiative: it slowly unravelled once its administrative champion left. Parents are another stakeholder group that need to be actively involved in the process of school food reform. The UK furnished an instructive example of what can go wrong if they are ignored when, in reaction to the Jamie Oliver television campaign, school meal take-up rates fell owing to the adverse reaction of discontented parents (BBC, 2006).

Financial investment is, of course, a vital ingredient in the recipe for a successful sustainable food procurement initiative. A pioneering initiative in East Ayrshire, Scotland, for example, was driven by the availability of funding from the Scottish Government's Hungry for Success initiative (Morgan and Sonnino, 2008). The Food for Life Partnership (FFLP) programme in England also owes much of its success to dedicated funding from the UK's Big Lottery fund (Orme *et al.*, 2011).

Another dimension of complexity is the highly contextual nature of the public plate. Standards and practices of public sector food procurement vary considerably from country to country and between localities in the same country. Within the UK for example, there are enormous differences in scale, structure, food quality and degree of sustainability across operations. Procurement systems can encompass multinational companies, SMEs, farmers and social enterprises in addition to the in-house catering facilities of public sector organisations. The scale can range from large multinational businesses such as Compass and Sodexho, which have a power of purchase equivalent to the major supermarkets, through large public sector buying consortia, down to individual primary schools buying direct from shops and farmers. Such diversity may become a thing of the past in the UK because one of the great questions facing the public sector catering system is whether the age of austerity privileges the low cost operations of the multinationals at the expense of more locally based suppliers such as SMEs and social enterprises, a question we explore later.

Similarly, within public sector institutions, buying food can be done centrally, by professional procurement officers, or devolved down to the individuals responsible for the catering service or event. This level of complexity presents problems both in terms of identifying representative systems and propagating institutional innovation and best practice. Within the UK we are beginning to see a growing institutional polarisation between England, which is devolving the power of purchase from the local authority level to the individual school, and Scotland and Wales, where local authorities are still the main institutional player in school catering. Within England, local authorities are also framing sustainable procurement policies in different ways, reflecting local needs and local political priorities.

Sandwell Metropolitan Borough Council, for example, a largely urban local authority in the West Midlands of England with one of the highest levels of social deprivation in the UK, has been a pioneer of healthy food procurement as a result of its partnership with the Sandwell Primary Care Trust. Sandwell represents an instructive counterpoint to more typical initiatives that tend to be located in more rural and often affluent areas. One of the key findings of a Sandwell research project was that the social and economic dimensions of sustainability were the priorities for stakeholders in the Borough rather than the environmental concerns that are more typically associated with sustainability. The framing of the sustainable procurement policy in Sandwell was such that it aimed to promote the interests of small businesses located within the borough, providing much needed employment opportunities and economic ripple effects within the local community (Davis and Morley, 2008).

Across the globe the factors that shape the power of purchase of food vary a great deal. In much of the developing world, for example, the focus has been on food access and security. Brazil, a country with varying types of school feeding programme since the 1950s, is one such country. Only recently have explicit provenance issues started to be integrated, most notably through the establishment in 2003 of the Food Acquisition Programme (Programa de Aquisição de Alimentos). More recent developments have included the decentralisation of school meal provision which enabled local level state institutions, service users and civic groups to participate in the design of such programmes (Otsuki, 2011). In recent years Ghana has been one of the leading African countries to experiment with the Home Grown School Feeding programme, which aims to produce a double dividend by providing healthy food for children, while sourcing it locally to provide markets for local producers. Once again, however, reformers underestimated the complexities of the Home Grown School Feeding programme, which is one of the reasons why progress has been slower than its UN sponsors anticipated (Morgan and Sonnino, 2008; Morgan *et al.*, 2007).

In many countries, particularly Northern Europe and North America, some of the most successful examples of sustainable food procurement have remained in discrete organisational boundaries with little spillover into proximate practices or adjacent organisations. In other words, they can be described as 'islands of good practice' (Morgan and Morley, 2003). While these initiatives are invaluable in research terms to understand how sustainable systems develop and are sustained (or not), they also lead to a focus on the prevailing mainstream system and provoke questions as to how such 'good practice' can inform and ultimately raise the average standard, so that good practice becomes the norm rather than the exception (Morgan, 2008).

Our research with the Food for Life Partnership explored this issue of good practice and why it was often such a 'bad traveller'. One of the principal aims of the partnership was to create 'flagships' of good practice (based on Food for Life criteria) and promote their diffusion both into neighbouring schools and within their own communities. Although not common, evidence was found of flagship schools raising sustainable food standards in neighbouring schools by changing the sourcing habits of shared caterers. For example, the presence of four flagship schools in Nottinghamshire encouraged the Local Authority caterer to provide FFLP standard school meals across its entire primary school clientele base. Similar examples exist with private contract caterers, who were encouraged to explore possibilities for turning FFLP into a mainstream service owing to the success of a handful of FFLP flagship client schools.

Despite these attempts to extend the practice of sustainable procurement, we often came to the conclusion that good practice was a bad traveller, which is another way of saying that the dissemination of innovative ideas and practices is far less common than neoclassical economic theory would have us believe (Morgan and Morley, 2006). Much of the evidence from international research would seem to suggest that truly embedded sustainable public plate systems are few and far between. In the following section we try to distil some of the barriers that have stymied the transition to a more sustainable public plate.

Values for money: the politics of the public plate

Public procurement professionals find themselves on the front line in the age of austerity because, while they are under pressure to secure 'more for less, they are also expected to deliver 'values for money' rather than just value for money in the conventional sense of the term (Lang, 2010; Morgan, 2012). In this final section we address the highly significant issue of *values-in-action*: that is to say the values that frame the narratives and policies that help us to harness the power of purchase. We begin with 'the ethical turn' in the social sciences as a prelude to discussing the role of values in public procurement policy. Beginning with the theoretical debate about values in social science is not as 'academic' as it might seem. On the contrary, the antipathy towards using values in the worlds of business and public policy has its roots in the prejudice against values in mainstream social science. This is a modern version of the problem that Keynes famously exposed when he said that 'practical men, who believe themselves to be quite exempt from any intellectual influences, are usually the slaves of some defunct economist' (Keynes, 1973).

Values within reason

Social science has a problem, according to social theorists such as Andrew Sayer, because it cannot convincingly explain why things matter to people (Sayer, 2011). This damning indictment is part of a critical realist critique that argues that the understanding of values in mainstream social science is seriously deficient because of the common assumption that values are beyond the scope of reason. As Sayer argues:

> In social science, it is common to regard values in emotivist or subjectivist terms, as not being about anything, except perhaps the holder's emotional state of mind. They are often seen as conventional – as merely derived from social norms – rather than as valuations of circumstances and actions. This ... has a detrimental effect both on social science's interpretations of social life, and on its own self-understanding. In the former case, it prevents social scientists from identifying why anything matters to people, and hence what kinds of things motivate them. In the latter case ... sociology and other social sciences have still not adequately come to terms with the *reason-laden – or reasonable – character of values*, so that

> there is still an aversion to normativity, that is to offering valuations of social phenomena, since values are seen as a source of bias and a threat to objective thought ... This weakens social science's ability to understand and convey why anything matters to actors, why values and norms have normative force, or why actors or researchers see anything as good or bad.
>
> (Sayer, 2011, p. 24)

The gist of Sayer's argument is that it is not values per se that pose a problem for social science, but dogmatically held values. Far from being beyond or opposed to reason, in other words, values are part and parcel of what moral philosophers refer to as 'ethical reasoning' and the 'intelligence of the emotions' (Nussbaum, 2001). Ethical reasoning challenges the widely held notion in mainstream social science that positive and normative statements must be inversely related, when in actual fact they can provide enriched explanations when used in concert. To illustrate the point Sayer refers to two different accounts of the Holocaust: 'thousands died in the Nazi concentration camps' as opposed to 'thousands were systematically exterminated in the Nazi concentration camps'. The latter, he argues, is both more value-laden *and* more factually accurate than the former, proving that evaluative terms can strengthen rather than weaken the truth of our explanations. The fact-value dualism has the effect of 'subjectivizing and de-rationalizing values so that they become "personal" rather than open to intersubjective deliberation and evidence' (Sayer, 2011, p. 45).

Sayer's realist critique of mainstream social science draws on the classical philosophies of Aristotle and Adam Smith as well as the contemporary ethic of care literature championed by feminist theorists such as Joan Tronto, all of which spans the divide between positive and normative analyses. What is most distinctive about the ethic of care literature, according to Sayer, is that 'it challenges the undersocialised and disembodied models of human being that have dominated political and moral philosophy' (Sayer, 2011, p. 257). This brings us back to the main theme of this chapter, which is how the public realm – through its power of purchase – can be harnessed to promote more sustainable outcomes in the context of public sector food provisioning. Recent research on public food procurement has also drawn on the ethic of care literature because it provides the most compelling ethical argument as to why the public realm needs to take more seriously its duty of care to vulnerable beings such as children who are being fed in public spaces such as school canteens (Morgan and Sonnino, 2008). The values of public health, social justice and ecological integrity – the core values of a capacious sustainable development paradigm – need to be factored into the calculations of public procurement managers and the first step in this direction is to actually recognise the existence of values other than low cost in the design of public tenders.

Recognising values

Far from being an arcane technical matter, the issue of metrics lies at the centre of the contested politics of the public plate. Metrics refers primarily to the core values of a public tender document. Covering cost and quality factors, they are weighted according to the priorities of the purchasing authority, with the contract awarded to the supplier that is most closely aligned with the metrics of the tender. As we have seen, cost has been the dominant metric for procuring the ingredients of the public plate in the UK, a country where low cost was allowed to masquerade as 'best value' during the era of Thatcherism. However, a *sustainable* public plate calls for a radically different kind of catering and procurement culture, where values for money is the goal and not value for money in the narrowest sense of the term. This means that the public realm has to recognise and integrate the core values of sustainability, namely the values of public health, social justice and ecological integrity. Academic research has helped to put this issue on the political agenda by arguing that adopting metrics of sustainability would involve internalising the costs that have been externalised in the conventional food system (e.g. the public health costs associated with diet-related disease and the environmental costs of intensive agri-food systems). Recognising these costs and getting them factored into the criteria for the award of a contract are two of the most difficult tasks facing analysts and advocates of the sustainable public plate because the quantification of health and environmental costs, a highly contested exercise even among economic experts, is beyond the normal capacity of hard-pressed public sector procurement managers. Even so, the key point to be established here is that the formal criteria for the award of public contracts in the EU have been broadened in recent years and rendered less restrictive, to the point where they now explicitly embrace some social and environmental factors. There is now little justification, therefore, for public managers to blame EU regulations if they continue to award contracts on the basis of low price (Morgan and Sonnino, 2008).

Measuring values

If the public plate is to be viewed and valued in a more sustainable way, analysts and advocates will need to be able to communicate the potential benefits of a sustainable public plate. Although sustainable food procurement is the epitome of good sense for consumers and citizens who set a high premium on the values of sustainability, politicians, policy makers and indeed the general public require more compelling evidence of these benefits.

Demonstrating the potential benefits beyond the direct economic impact is, of course, one of the most important (and contentious) challenges in every sector of the sustainability debate. The social return on investment (SROI) methodology, for instance, has emerged in recent years as a significant tool for converting complex socially based impacts into a single economic value (Lawlor *et al.*, 2008). However, its inherent complexity renders it unsuitable for widespread adoption as a systematic and practical tool to measure impact within individual procuring organisations. Similarly, the concept of 'whole life costs' has been hotly debated among advocates of sustainable purchasing, a concept that requires a good deal of time and effort because many rival products have to be evaluated at the same time. Life cycle analysis (LCA) is one of the most popular methodologies for assessing whole life costs and can produce some very revealing and therefore powerful results regarding the long-term balance of costs and benefits (e.g. Tarantini *et al.*, 2011). All these methodologies involve a trade-off between depth and utility in 'everyday' situations that restrict their routine adoption. The time-consuming process of quantifying environmental and social benefit appears to be too costly for cash-strapped public sector organisations to do on a regular basis.

Indeed, the systematic recording of basic data – regarding what is purchased and from where etc. – appears to be problematic for many public sector organisations and their suppliers. The evaluation of the FFLP illustrated a congenital inability to record and evaluate basic data at all levels of school food provisioning in England. During the course of this research project, the academic evaluators and the FFLP parties had great difficulty in gaining access to supply chain data from caterers, particularly with respect to financial data. Despite concerted efforts to secure key data, many caterers were either unwilling or unable to provide the necessary information on costs and benefits. To some extent this was simply a reflection of the poor data management systems typically used by public sector caterers. But it may also have been a reflection of the low priority attached to gathering evidence of impact by both caterers and FFLP as a whole. In-house and private single-site caterers may consider themselves as too small to need systemised sourcing data collection systems, while large local authority or national contract caterers are so big that they agglomerate their sourcing data. In both cases, it appeared difficult for caterers to provide the appropriate information unless they had a direct incentive to do so (Orme *et al.*, 2011).

Generally speaking, this 'knowledge deficit' similarly reflects a low status attached to provenance issues across the conventional food system, particularly among non-animal production supply chains, which have legal requirements for traceability (though this is no guarantee of provenance, as the horsemeat scandal demonstrates). What the knowledge deficit also illustrates is the fact that robust monitoring and evaluation systems are conspicuous by their absence in the public sector and that this constitutes a major organisational impediment to data collection, without which it is impossible to measure the costs and benefits of the public plate that stakeholders demand.

Enabling values

Enabling sustainable values to be realised in practice – through the application of professional skills, new business models and political commitment – is the most important aspect of the power of purchase, an aspect that raises probing questions about the competence and confidence of the public sector to act as an agent of change. The three dimensions identified above – the level and mix of professional skills in the public sector, the need for more innovative business models and the politics of the public realm – are arguably the key challenges for the public sector to act as a credible agent of change and a champion for a more sustainable food system.

The level of professional skills in the public sector is an issue that appeared with depressing regularity during our public plate research and this skills deficit was especially pronounced with respect to public procurement. A lack of sustainable procurement skills helps to explain why the public sector systematically fails to apply whole life costing to its purchasing decisions. In 2006, the Sustainable Procurement Task Force argued that 'many parts of the public sector currently lack professional procurement expertise and that people are routinely allowed to spend money without being appropriately trained' (SPTF, 2006, p. 47). This skills deficit chimes with an earlier inquiry which found that less than a quarter of all procurement staff had the requisite professional skills (NAO, 2005; Morgan, 2010).

Beyond professional skill sets, there are the equally important but less tangible issues of confidence, leadership, risk management and organisational innovation, all of which are the hallmarks of the few not the many (Morgan and Morley, 2006; Morgan and Sonnino, 2008). In Wales the procurement skills deficit has reached crisis proportions, which means that the public sector will find it impossible to deploy its power of purchase without a major investment in targeted skill formation (Morgan, 2010).

New business models – such as food hubs, social enterprises and public–social partnerships – also need more attention from analysts and advocates as vehicles to introduce new values into the food system and forge more sustainable links between producers and consumers. The key issue for all business models, however, is financial sustainability, the precondition for all forms of sustainability. Forced to make deep public expenditure cuts, local authorities in and beyond the UK feel obliged to outsource services that were hitherto designed and delivered by their own public sector service departments. Many of these public services – such as the school meals service for example – are being outsourced to large private sector service companies, such as Compass and Sodexho, who are seeking to 'bundle' welfare services with IT and cleaning services as part of a larger 'facilities management' contract with local government clients. One of the reasons why the private sector is deemed to be the only alternative to public sector provision is because the public sector has not invested enough time and energy into framing new forms of enterprise – such as civic enterprise, social enterprise and public–social partnerships (Morgan and Price, 2011). One of the most imaginative civic or social enterprises to have emerged in recent years is Whole School Meals Ltd (WSM), a highly innovative business model because it is predicated not on competing with the public sector but on partnering it. Indeed, WSM designed its business model in such a way to enable state sector schools to be the shareholders of the company that was supplying high-quality and locally produced food for local schools in the county of Kent, England.

As a community-based social enterprise, WSM is a potential model for other areas, providing a new imaginary for sustainable businesses and an organisational alternative to the outsourcing of services to private sector companies whose principal loyalty is to distant shareholders rather than to local people as is the case with WSM. Created as a social enterprise in 2005 by a group of parents, school governors and local entrepreneurs, 75 per cent of the shares in the business are owned by the eighteen schools it serves in Kent. The organisation aims to be commercially oriented and seeks to generate a profit where possible, which is then either reinvested back into the business or returned to the schools themselves. WSM was set up with the aim of providing direct community benefits as well as having children's well-being at heart. This latter goal is met by producing meals that are nutritious and appealing to children using fresh and seasonal ingredients. Produce is sourced from local producers whenever possible.

Box 4.2: Whole School Meals Ltd mission statement

- To provide nutritious, tasty food for children which supports their health and well-being
- To source local produce whenever possible
- To maintain the highest professional standards including food hygiene and health and safety
- To work closely with the local community of schools, parents and children and consult with them about our activities and key decisions
- To value every member of staff and seek to reward staff fairly and involve them in the Company's development
- To be commercially viable and financially independent

Source: www.wholeschoolmeals.co.uk

This social enterprise model allows schools and other stakeholders to retain control over their school meals service without having to bear the burden of day-to-day management. By working with local entrepreneurs, the public sector is able to focus on what it does best, while devolving the commercial aspects to a social enterprise framework that provides legal and moral reassurances for the schools themselves and their stakeholders.

Social enterprise models such as WSM usually rely on *ad hoc* grants and other public funding arrangements to be sustained. If the power of purchase is to be fully realised through public–social partnerships, then funding mechanisms need to be broadened and secured in the longer term. In the UK, there can be no doubt that the contribution of National Lottery funds has been a key factor behind the successful launch of many sustainable food initiatives. Launching such sustainable food initiatives is one thing, but assisting them to become financially sustainable is perhaps more important, which suggests that far more attention needs to be paid to developing new community-based models of enterprise (Morgan and Price, 2011).

Of course, one of the reasons why the social enterprise sector continues to rely on public sources of funding is because the dominant form of economic calculation in capitalist society fails to adequately account for the full social and environmental costs associated with production, distribution and consumption. Hopefully, proposed changes to EU procurement rules, under negotiation at time of press, will include the ability for government organisations to include 'whole-life' costs, and therefore social and environmental values, when they are finalised (Morgan, 2012). This would enable public authorities across Europe to design and deliver services through social enterprise and other forms of public–social partnership to make it more likely that the societal benefits of their actions are maximised at a local level, where communities can more readily scrutinise the results.

While professional skills and social business models can help the public sector to become a more effective agent of change, they count for little in the absence of the political will to harness the power of purchase for more sustainable ends. This is especially important today because of the confluence of a food crisis and the age of austerity, a toxic amalgam that could nullify all the 'little victories' that have been painstakingly secured by school food reformers over the past decade. The recent food crisis triggered by the discovery of horsemeat in processed meat products labelled as beef embroiled public sector caterers as much as powerful supermarkets. Indeed, the most important lesson of the crisis is that mainstream food chains have become so long, complex and convoluted, straddling so many tiers and transactions, that even the most powerful players cannot guarantee the integrity of their supply chain. If powerful players such as Nestlé, Walmart and Tesco are unable to monitor and regulate their supply chains, how can a cash-strapped local authority caterer do so?

To rebuild trust in the integrity of the public plate will take time, resources and the political will to use the power of purchase to champion the cause of shorter and more transparent food chains in which a higher premium is attached to provenance. Although shorter food chains ought to be more easily monitored and regulated, because the transactions are fewer and more localised, the local label is no guarantee of integrity. A tragic case in point was the E.coli crisis in Wales in 2005, which claimed over 150 victims and the death of a five-year-old, which was caused by a local butcher who was convicted of supplying unsafe meat to forty-four schools in the South Wales Valleys.

If the public sector is to become a more effective agent of change in the food system, the status of the procurement function needs to be enhanced so that it is conceived as a strategic instrument to refashion markets, production and consumption. This is a process that requires new and improved skills within the public sector and, increasingly, new business models in civil society to provide alternatives to the outsourcing of public services to the private sector (Morley *et al*., 2012).

Whether a viable school food service will survive the age of austerity in the UK, with its savage and counterproductive cuts in public expenditure, will depend on the way the service is viewed and valued. If it is viewed and valued in narrow commercial terms, where profit and loss are the only metrics, then the service looks doomed to decline, possibly dwindling into a statutory rump of a highly stigmatised free school meals service, the preserve of the poorest of the poor. But if it is viewed and valued in more capacious terms, where public health, social justice and ecological integrity are the key metrics, then it is possible that a future school meals service will be recognised for what it could be and should be – a health-promoting service for all.

References

Barlett, P.F. (2011) 'Campus sustainable food projects: Critique and engagement', *American Anthropologist*, vol. 113, pp. 101–115.

BBC (2006) 'Fewer pupils eating school meals', BBC News Online, 13 July, available at: news.bbc.co.uk/1/hi/education/5177880.stm, accessed 11 May 2013.

Belot, M. and James, J. (2009) *Healthy School Meals and Education Outcomes*, Oxford: Centre for Experimental Social Sciences, Nuffield College.

Davis, L. and Morley, A. (2008) 'Eatwell in Sandwell' sustainable food procurement project, research report for Sandwell Metropolitan Borough Council, Cardiff: Ital Associates/Cardiff University.

Defra (2012) *Food Statistics Pocketbook 2012*, Department of Environment, Food and Rural Affairs, London: HSMO.

Finkelstein, E.A., Trogdon, J.G., Cohen, J.W. and Dietz, W. (2009) 'Annual medical spending attributable to obesity: Payer and service specific estimates', *Health Affairs*, vol. 28, no. 5, pp. 822–831.

Izumi, B.T., Wynne Wright, D. and Hamm, M.W. (2010) 'Market diversification and social benefits: Motivations of farmers participating in farm to school programs', *Journal of Rural Studies*, vol. 26, no. 4, pp. 374–382.

Keynes (1973 [1936]) *The General Theory of Employment, Interest, and Money: The collected writings of John Maynard Keynes, Vol. VII*, London: Macmillan Press.

Lang, T. (2010) 'From "value-for-money" to "values-for-money"? Ethical food and policy in Europe', *Environment and Planning A*, vol. 42, pp. 1814–1832.

Lawlor, E., Nicholls, J. and Nietzert, E. (2008) *Measuring Value: A guide to social return on investment*, London: New Economics Foundation.

Meadowcroft, J. (2007) 'Who is in charge here? Governance for sustainable development in a complex world', *Journal of Environmental Policy and Planning*, vol. 9, no. 3, pp. 299–314.

Morgan, K. (2008) 'Greening the realm: Sustainable food chains and the public plate', *Regional Studies*, vol. 42, no. 9, pp. 1237–1250.

Morgan, K. (2010) 'Local and green, global and fair: The ethical foodscape and the politics of care', *Environment and Planning A*, vol. 42, no. 8, pp. 1852–1867.

Morgan, K. (2012) 'Values for money', *Agenda*, Spring.

Morgan, K. (2014) 'The new urban foodscape: Planning, politics and power', in A. Viljoen and K. Bohn (eds) *Second Nature Urban Agriculture: Designing productive cities*, London: Routledge.

Morgan, K. and Morley, A. (2002) 'Re-localising the food chain: The role of creative public procurement', Cardiff: The Regeneration Institute, Cardiff University.

Morgan, K. and Morley, A. (2003) 'School meals: Healthy eating and sustainable food chains', Cardiff: The Regeneration Institute, Cardiff University.

Morgan, K. and Morley, A. (2006) 'Sustainable public procurement: From good intentions to good practice', Cardiff: The Regeneration Institute, Cardiff University.

Morgan, K. and Price, A. (2011) *The Collective Entrepreneur: Social enterprise and the smartside*, Project Report, Cardiff: Charity Bank and Community Housing Cymru Group.

Morgan, K. and Sonnino, R. (2008) *The School Food Revolution: Public food and the challenge of sustainable development*, London: Earthscan.

Morgan, K.J., Bastia, T. and Kanemasu, Y. (2007) 'Home grown: The new era of school feeding', *Project Report*, Rome: World Food Programme.

Morley, A., Sonnino, R. and Smith, A. (2012) 'Leading by procuring: The power of public sector purchasing', in D. Rigling Gallagher (ed.) *Environmental Leadership: A reference handbook*, Thousand Oaks, CA: SAGE Publications, pp. 113–122.

NAO (2005) *Sustainable Procurement in Central Government*, London: National Audit Office.

Nussbaum, M. (2001) *Upheavals of Thought: The intelligence of emotions*, New York: Cambridge University Press.

OECD (2002) *The Size of Government Procurement Markets*, Paris: Organisation for Economic Cooperation and Development Publishing.

OECD (2010) *Obesity and the Economics of Prevention: Fit not fat*, Paris: Organisation for Economic Cooperation and Development Publishing.

de Onis, M., Blössner, M. and Borghi, E. (2010) 'Global prevalence and trends of overweight and obesity among preschool children', *The American Journal of Clinical Nutrition*, vol. 92, no. 5, pp. 1257–1264.

Orme, J., Jones, M., Kimberlee, R., Weitkemp, E., Salmon, D., Dailami, N., Morley, A. and Morgan, K. (2011) *Food for Life Partnership Evaluation: Full report*, Cardiff: University of the West of England, Bristol/BRASS.

Orrey, J. (2003) *The Dinner Lady*, London: Transworld.

Otsuki, K. (2011) 'Sustainable partnerships for a green economy: A case study of public procurement for home-grown school feeding', *Natural Resources Forum*, vol. 35, no. 3, pp. 213–222.

Sayer, A. (2011) *Why Things Matter to People: Social science, values and ethical life*, Cambridge: Cambridge University Press.

School Food Trust (2012) *Statistical Release: Take up of school lunches in England 2011–2012*, Sheffield: School Food Trust.

Scottish Executive (2002) *Hungry for Success*, Edinburgh: Scottish Executive.

SPTF (2006) *Procuring the Future: Sustainable procurement national action plan*, Sustainable Procurement Task Force, London: Defra.

Tarantini, M., Loprieno, A.D. and Porta, P.L. (2011) 'A life cycle approach to green public procurement of building materials and elements: A case study on windows', *Energy*, vol. 36, no. 5, pp. 2473–2482.

WHO (2012) *World Health Statistics 2012*, Geneva: World Health Organization.

5 Sustainable food supply chains

The dynamics for change

Andrew Flynn and Kate Bailey

Introduction

Understanding the dynamics of food and sustainability are pressing academic and public policy concerns. The food system in developed countries, especially that of the UK, often seems to be bedevilled by periodic alarms, for example, about the safety of food or its nutritional value. Anxieties about food safety and quality are allied to broader concerns about the relationships between producers and retailers, between retailers and manufacturers, and between economic actors in the food chain and policy makers and regulators.

While public policy concerns relating to food are long-standing, contemporary debates on food and sustainability appear to have their own distinctive features. First, they involve a wider range of actors: all players in the supply chain have a legitimate voice in raising issues about how sustainability should be thought about and acted upon, and so too do a variety of other actors, ranging from bodies concerned with trade, alternative food networks, health, the environment and so on. In short, the public policy arena for dealing with food issues has become more crowded. Second, and partly as a consequence of the plethora of voices clamouring to be heard on food and sustainability, there is no fixed or agreed agenda. Many issues appear to wax and wane with regularity as their promoters bring them to the fore only to see them rapidly replaced by another issue, while others appear to have greater longevity, such those linked to climate change. Issues, however, do not simply disappear; rather they help to inform the way in which subsequent issues emerge. Policies are layered one upon another (Feindt and Flynn, 2009); issues interact, coalesce and compete in highly dynamic ways. Growing awareness of global threats to the food system has contributed to these debates, adding a further layer as concerns of food security and those of ecological sustainability converge (Godfray *et al.*, 2010; SDC, 2009). This has prompted calls for new agri-food systems that recognise the multi-functionality of food; its social and environmental contributions in addition to its economic ones (IAASTD, 2009; House of Commons, 2009; Foresight, 2011). Advocates argue that new food systems need to be shaped to withstand greater volatility and uncertainty (Ingram *et al.*, 2010; Foresight, 2011) and need to operate within ecological limits, remain competitive while delivering fairer returns and greater social benefit (Pretty, 2008; Ambler-Edwards *et al.*, 2009; The Royal Society, 2009; Garnett and Godfray, 2012). So, in recent years there have been, and continue to be, debates on food miles, food and health, food and climate change, the environmental impacts of food products, food and trade, alternative food systems, and food security.

In this chapter we ask, to what extent is the current dominant food system capable of internal transformation to make it more sustainable? What is the potential for, perhaps, the key alternative food system, organic production, to engender change in the broader food system? The chapter is organised around five themes: the first discusses food supply chains in a global context; second, what is meant by food systems; third, the nature of alternative food systems; fourth, the potential transformative capacity of organic food; fifth, the environmental impacts of the food system, and concludes by pointing to some of the pressing academic and policy concerns.

Food supply chains and globalisation

Since the mid 1980s, there has been increasing recognition that the state is withdrawing from agrarian markets, and that international economic actors, especially retailers, seek highly flexible supply chains to meet the increasingly differentiated demands of consumers in the developed world. Global food systems have national or sub-national impacts. For example, a number of commentators have highlighted the role of retailers in restructuring and directing food supply chains (see, for instance, Marsden *et al.*, 2000; Hendrickson and Heffernan, 2002; Fulponi, 2006; Fuchs *et al.*, 2011; Havinga, 2012). As American commentators have noted, however, the increasing dominance of food retailers is newer to the USA than it is to Europe (Hendrickson and Heffernan, 2002, p. 357; Guptil and Wilkins, 2002, p. 40) where the Dutch and UK markets, for example, have for some time been dominated by a small number of highly influential supermarkets (see Oosterveer, 2012). The most influential retailers are likely to be European-based transnationals, such as Tesco (UK), Metro and Lidl (Germany) and Carrefour (France). Walmart is likely to be the only major US-based global retailer. Although the US has relatively recently witnessed the levels of dominance of a small number of key players that have been found for some time in countries such as the UK and the Netherlands, the results are similar: retailers dictate terms to food manufacturers who then force changes back through the food system (Hendrickson *et al.*, 2001; Konefal *et al.*, 2005). While it is tempting to point to an almost inescapable logic of dominance by retailers of the food system in the developed world, to do so would ignore the very real regulatory, social and economic factors that can give local and national food systems their own distinctive characteristics.

Another feature of European retailer-led food systems is that many also have a more marked trend towards own label brands than is to be found in the US. In the UK, for example, in 2011 sales of retailers own label goods were greater than those of branded items (Mintel, 2012). In the US, meanwhile, private label brands have been more successful at defending their position. For example, Kroger estimates that its own label accounts for 25 per cent of sales in the early 2000s (Hendrickson *et al.*, 2001, p. 13) and that for private brands more generally had only risen to the high 20s a decade later (*Wall Street Journal*, 2012). The prominence of retailers in the UK food system and the increasing identification of branding with a supermarket rather than a product is important for the ways in which the food supply chain has operated in practice and how it has conceptualised environmental sustainability (see the example of Sainsbury's below). In part, the difference between some major European retailers and their American counterparts arises from the way in which the former have sought to identify themselves as guardians of the consumer interest. Consumers trust in own label brands is high. Meanwhile, the latter can often be viewed as brokers who procure products at best value for consumers, so resulting in a narrower, more transactional relationship between consumer and retailer.

At the heart of the conventional understanding of the supply chain is the buyer–supplier relationship. The primary objective is to minimise production and transaction costs. Ring and Van de Ven (1992) argued that buyer–supplier relations could be divided into three types dependent upon the levels of trust and risk in the relationship. Low levels of risk and/or little need for trust are best governed by markets, and this may well have typified traditional relationships within the food supply chain. High levels of trust and low levels of risk are best managed by recurrent contracts. High levels of risk and low levels of trust require hierarchies. Food safety and sustainability are prominent public policy issues; this suggests that there will be high levels of both trust and risk in the supply chain and if these are to be successfully managed demands closer supplier–buyer relations.

The concept of *integrated supply chain management*, as identified by Storey *et al.* (2006), is founded on the premise that integration, or increased coordination of processes and activities, between actors and organisations within a chain (be it two or more echelons) creates increased efficiencies and higher levels of competitive advantage (Smith, 2008). In the same vein the ability to align customer value propositions and focus also leads to greater competitive advantage for the members. This assumes a level of collaboration between supply chain actors generating mutual benefit. As Bloom and Hinrichs (2010) point out, though, practice demonstrates how difficult it can be to put together food distribution networks that meet the needs of both producers and consumers.

In the UK agri-food industry, retailer-based collaborative partnerships have been highlighted as often one-sided with a low level of mutuality (Ireland and Bruce, 2000; Cox and Chicksand, 2005). The largest four retailers operate as an oligopoly, effectively controlling large parts of the market.

In spite of this, the market has been judged to be 'working' in that it delivers consumer benefit in terms of price, product quality and range (Competition Commission, 2000), despite the power imbalance exhibited between retailers and the rest of the supply chain (Hingley, 2005). However, it is now more accepted that there is a level of unfairness in the system and some measures have been introduced to curb this, through the introduction of the Code of Practice for retailers (the Groceries Supply Code of Practice) in 2010 and an independent Groceries Code Adjudicator established in 2013. To what extent though these will address the structural issues of unequal power relations at the heart of the system is doubtful. For example, Cox *et al.* (2007) and Francis *et al.* (2008) in their detailed studies of red meat supply chains in the UK show that the multiple retailers and to a lesser extent integrated processors can dominate supply chains through their buying power. This means that governmental strategies based on voluntary means or the promotion of collaboration in the supply chain are likely to be inadequate because they will not sufficiently challenge existing power relationships and the Code of Practice, while promoting the idea of fairness in relationships, will undoubtedly not challenge commercial realities. Until relationships are reconfigured, Cox *et al.* (2007, p. 690) argue that rather than 'promoting sustainable competitive advantage for all, the outcome that is most likely for the majority of participants [in red meat supply chains] is a commercial "treadmill to oblivion" of continuous operational improvement, with low and declining returns'.

In his study of Sainsbury's environmental supply chain management, Hall (2000, p. 463) pointed out:

> Compared to the total environmental impacts generated by their supply chain, Sainsbury's legal responsibilities was small and easily managed. However, this did not stop them from having an impact over their suppliers' environmental activities, especially where they had legitimate reasons to influence them, such as for own-brand products.

Sainsbury's interventions in the supply chain took particular forms, for example, providing guidance on pesticide reduction and animal husbandry. Hall (2000) detects signs of improved environmental supply chain management, but notes that '[t]his was restricted to high-profile issues which drew the attention of environmental advocacy groups, such as animal rights, pesticide use, sustainable forestry and the elimination of CFCs, and targeted at Sainsbury's' (Hall, 2000, p. 464).

While the early forays of supermarkets into environmental supply chain management may have been somewhat limited and marked by an issue-based approach, they have deepened and widened over time (see, for example, the work of Spense and Bourlakis (2009) on Waitrose). In the UK, all the major retailers have commitments to reduce carbon within their own stores and direct logistics operations (e.g. 'green stores', green distribution centres, transport to store, etc.).This is an area that has seen continued focus since the mid 2000s, for example Marks & Spencer's operations and distribution achieved carbon neutral status in 2012 (Marks & Spencer, 2012). The reduction of waste has also been a keen focus area, driven in part by high landfill charges, and all retailers have zero landfill targets. Recent developments have seen moves to reuse waste as energy; Sainsbury's, in particular, have invested in anaerobic digestion technology (Sainsbury's, 2012). Sustainable sourcing of raw materials such as wood and palm oil figure highly, along with 100 per cent procurement of sustainable sources of fish. Concerns over efficient resource utilisation have started to filter through, for example Sainsbury's and Marks & Spencer are working with overseas suppliers in areas of water scarcity to support reduction in usage. In addition, local sourcing has become an established feature in all the major supermarkets. However, where organic foods did once feature, it is noticeable there are no longer any specific targets set by the retailers to invest or source these products; for example, Tesco and Iceland had targets that were subsequently dropped. Health concerns and the rise of obesity have triggered some change in approach. For example, there has been a noticeable rise in product reformulations, primarily to reduce fat, sugar and salt, particularly in own brand products. Change, though, is reformist rather than radical, working with the grain of supermarket strategies rather than providing a challenge to them.

Concerns over price volatility and the availability of supply, along with consumer interests in provenance, have driven retailers to forge more direct relationships with farmers. The last few years has seen the establishment of more direct retailer–producer groups (Marks & Spencer, 2010; Morrisons, 2010). Retailers have started to work proactively with some of their farming base to promote more sustainable practices such as a reduction in pesticides, emission controls in dairy/meat farming, improvement in animal welfare standards and increased traceability from farm to fork. For example, Morrisons made much of its farm to shop integrated supply chain to reassure its consumers during the horsemeat in the supply chain scandal in early 2013. Biodiversity, an often forgotten area of sustainability, has started to attract attention, with the introduction of small pilot projects to promote biodiversity on selected farms by Tesco and Marks & Spencer, among others. Despite this, there still remain huge swathes of the food chain, particularly the frozen and processed foods channels, along with much of the general farming base that are not included in these types of initiatives. There are also noticeable gaps in commodities, particularly cereals which in effect take up 50 per cent of the UK's arable land (Defra, 2009). And while retailers have driven much waste out of their immediate operations, it is difficult to see whether systemic waste throughout the system (e.g. product rejected owing to aesthetic reasons) has been addressed. For the most part, and Morrisons may be something of an exception, there is still significant disconnection between the retailers and the farming base, which has meant that issues such as efficient resource use, soil management and biodiversity are, arguably, still not being sufficiently addressed.

These UK retail strategies indicate that the current direction of travel for the food system is focused on incremental, technology driven change (an adaptation of the food system) rather than innovative and radical restructuring, or system level change. The process is one of change and adaptation through product (e.g. reformulations) and process innovation (e.g. greening of intensive farm practices). Although there are some examples of radical system and structural change, these come in the form of alternative food networks (Sonnino and Marsden, 2006) and in local government and community food-based policies, aimed at the restructuring of food supply (see Food Futures Partnership, 2007; Morgan and Morley, Chapter 4, and Sonnino and Spayde, Chapter 9, both this volume). Typical alternative sustainable approaches to agriculture are characterised by organic, conservation and other ecologically based systems. However, all of these innovations exist in lower level technological niches and have not yet been able to demonstrate scalability. To explore the reasons why alternative and potentially more sustainable food systems do not 'break through' we outline the nature of the food system.

Food supply chains and food systems

The 'food system' can be conceptualised as a sequence of activities, starting with the production of plant seed, eggs or newborn animals. The sequence continues from production through intermediate processing and various distribution steps to generic retail or food service. The system ends with the consumption of food products by individuals. Since the system encompasses both production and consumption and examines inputs and outputs as well as the regulatory and policy context, it has been termed a food production and consumption system (FPCS) (Green *et al.*, 2003). Evaluations of the environmental sustainability of production consumption systems have, however, conventionally adopted an approach to setting boundaries based on the philosophy of considering products from 'cradle to grave', an approach that is now embodied in international standards for environmental life cycle assessments (Curran, 1996). Many studies of food products have subsequently followed this convention (e.g. Jungbluth *et al.*, 2000; Matos and Hall, 2007).

Life cycle assessment methodologies further suggest that a system level approach may be characterised by its inputs and outputs and that the type and magnitude of these flows is related to the impact of the system on its surroundings. While the flows of concern from an environmental standpoint are flows of substances, a more complete understanding should also consider less tangible flows such as those of knowledge, labour and financial subsidy that are of relevance to the overall social, environmental and economic sustainability of the system. It is important to note that innovations within the system can lead both to changes in the total value of particular flows (for instance when a lighter grade of packaging film is used to contain a given amount of product, so that the total flow of polymer into the system is reduced) and in the location of flows within the system (for instance when consumption shifts from home-prepared food to ready-meals, moving some labour input from the domestic realm (normally taken to be part of consumption) into the processing part of the system).

The potential for trade-offs between different actors, coupled with the importance in sustainability discourse of notions such as the equitable distribution of economic returns, lead us to the conclusion that it is reasonable to consider sustainability at this system level. Taking the system as the unit of analysis has a further benefit of allowing consistency between considerations of relative sustainability, considerations of consumption patterns, which Southerton *et al.* (2004) suggest are very much structured by the relations between key actors that enable them, and considerations of innovation, which Coombs *et al.* (2003) argue to be a phenomenon best seen as distributed across systems.

In analysing the stability (in the sense of reproducibility) of the dominant FPCSs and the potential for transformation, Transition Theory, the framework developed by Frank Geels and Johan Schot (2007) in which they develop a comprehensive framework to explore 'systematically how different kinds of multi-level interactions lead to different transition pathways' is particularly helpful. There are two key features to their model that are useful here: the first is the way it seeks to understand system transition as a process arising out of change at multiple levels, which is helpful given the complexity of current food systems. The second is that they posit five 'transition pathways', that is ways in which existing socio-technical systems (or 'regimes' as they call them) can be transformed.

Geels and Schot identify transition pathways that emerge out of interactions that occur at three different but interrelated levels. These are the 'meso'-level that encompasses the socio-technical regime; the micro-level where novelties ('technological niches') emerge; and the macro-level formed by the socio-technical landscape and consisting of factors outside of the control of niche and regime actors – trade negotiations would be an example – but that will influence the behaviour of key actors. For Geels and Schot (2007, p. 4) the core idea of their multilevel model is:

> [T]ransitions come about when processes at different levels link up and reinforce each other: a) the niche-innovation builds up internal momentum ... b) changes at the landscape level that create pressure on the regime and stimulate niche innovations, c) destabilisation of the regime, which creates a window of opportunity for niche-innovations. The alignment of these processes enables the breakthrough of the new configuration in mainstream markets where it competes with the existing regime. If the new configuration becomes dominant, it results in a socio-technical regime change.

The five transition pathways that Geels and Schot (2007, p. 11) identify are based upon different compositions of interactions at the three levels of their model. They term these transition pathways: 1) transformation, 2) opening up a new domain, 3) technological substitution, 4) de-alignment and re-alignment, and 5) reconfiguration. By pointing to the different ways in which transitions can arise, Geels and Schot provide a marker against which to test the potential for mainstreaming organic food. In particular, we can begin to explore why alternative food networks strike such a resonant chord with a number of producers and consumers, and then consider how the conventional food system interacts with the organic food system. It is this interaction that is one of the key insights offered by Geels and Schot, because transitions, and the potential for transitions, arise from the way in which niches engage with the mainstream and the mainstream engages with niches.

Alternative food systems

Alternative food systems seek to 'produce change in the "modes of connectivity" between the production and consumption of food, generally through reconnecting food to the social, cultural and environmental context of its production' (Kirwan, 2004, p. 395). As Kirwan points out, what counts as different (or otherness) when compared to the conventional food system will vary. For fair trade products, this is the promotion of equitable social relations between producer and consumer, despite the fact that much fair trade produce moves through, and is purchased from, the conventional system. For those who produce and consume organic food the sense of difference is drawn from valuing health, environmental or animal welfare issues. Among those who participate in alternative food systems there is a sense that they are producing and consuming foods that are of a different (higher) quality, whether because of localness or craft base for example, compared to that of the homogenised, conventional food system (Kirwan, 2004).

The creation of alternative food supply chains has the potential to alter the course of the dominant conventional food supply chain towards greater sustainability. This is because these alternative chains have different impacts in relation to space, time and structure of the food supply chain that typically have positive implications for sustainability. For example, within the dimension of space, the local food supply chain is smaller, which may have a positive local economic impact on local and national food producers and processors. Furthermore, transport costs are reduced and energy resources, when comparing like systems, are saved. Since food has to travel less, the local food supply chain may also operate faster. A local food supply chain will strengthen the structure of local business interactions and should favour small- and medium-size businesses, therefore enhancing the diversity of the chain. Overall, a local food supply chain could have a positive sustainability impact (Bellows and Hamm, 2001; Garnett, 2003; Morris and Kirwan, 2011). It is important, though, to extend a note of caution as, for individual foods, local food production may not be the best option (see Edward-Jones *et al.*, 2008).

In both the North American and European literatures there is now a wealth of case studies on alternative, more localised food systems. For instance, Hendrickson and Heffernan (2002, p. 362) write of the Kansas City Food Circle as an 'attempt to create a local, organic food system where consumers can get seasonal, fresh food at a price that supports farmers using sustainable practices'. In the UK, Kirwan (2004), for instance, has undertaken a detailed analysis of farmers' markets. Some of the most interesting work on alternative food systems has been that carried out by Murdoch and Miele (2004) and Marsden (2004). For Murdoch and Miele the emergence of alternatives are an 'artisanal reaction' and for Marsden a 'counter movement' to the dominant conventional food system. What the authors clearly demonstrate is the variety of alternative food movements; for Murdoch and Miele this is achieved through case studies of Slow Food, organics and fair trade, and for Marsden by reporting on the results of a seven country European study. As Tregear (2011) has helpfully pointed out in her review of current debates on alternative food networks there does need to be some caution in interpreting the benefits of non-conventional food supply chains as there is much confusion over the meaning of terms and of the claims that are being made on its behalf.

Organic food and system change

In theory it would be possible to construct an alternative, organic, food system to challenge the current conventional system. Agricultural production is, perhaps, the easiest stage to conceptualise the system differences since organic agriculture has to be certified to show that there is an absence of defined substances in the production process. Many would argue that organic production, since it promotes itself as a healthier and more environmentally friendly alternative, is likely to be associated with a reduction in the amount of energy used in the manufacturing, retailing and consumption of food. These values may also make themselves felt in a reluctance to engage with conventional manufacturing and retailing companies and to promote the virtues of seasonality and product variability since they arise from a 'natural' production process. In this section we assess the potential of organic chicken and potato production with respect to sustainable consumption. The following case study describes how a chicken producer moved from conventional to organic production, illustrating some of the differences and overlaps between the two systems.

Moving to an organic system: the limits of change

The farm business, one of the largest producers of organic chickens in the UK, moved into organic production in 1998 at a time when organic chicken production was highly fragmented and marginal. For some years prior to the late 1990s the farm had reared conventional poultry on a small scale, but in 1997 the farm experienced a severe financial crisis owing to the loss of key customers. This was a defining moment for the business as they could have either continued on the conventional route or move into a new niche. They chose the latter as they recognised they could not compete with the large poultry producers and, on a positive note, believed that some British consumers were seeking to change their shopping habits and become 'careful' consumers. In the beginning, the farm encountered the problem of procuring chicks suitable for organic production. Hatcheries in the UK were unable to supply the slow growing breeds that were required, so the farm eventually started sourcing from France, later bringing the production to 4,000–5,000 broilers a week.

What is notable about the shift in production system is that it was made for commercial reasons. Despite the strong normative element within the organic movement such values played no part in the decision making of the owners. Moreover, they did not notice a lot of change in moving from conventional to organic production. As one member of the family commented:

> The only changes are in the use of the food for indoor brooding. Indoor rearing is almost the same, apart from the feed. For outdoor rearing [a requirement of organic poultry production] it is just having the courage to open the doors. And we cannot rely on antibiotics [another condition of organic production].

To 'open the doors' for the chickens, so that they could engage with the environment outside their shed, is a significant phrase with a layered set of meanings. It implies a markedly different approach to rearing chickens, moving from a highly controlled environment to one in which the farmers become more dependent on understanding how their chickens interact, and flourish, within a more 'natural' and less controllable setting. At another level, the phrase is also about the family opening their minds, being open to innovative practices.

In reality, typically it is only at the primary production stage that the organic supply chain differs from conventional supply chain, and even here the differences, at least for one set of producers are not perceived to be radical. At the rearing stage there are distinct differences in breed, feed, period of growing and manner of keeping chickens. However, at the stages of food processing and distribution the two food systems are almost identical. In the processing stage, although slaughter houses should be certified as organic, the equipment is identical to that in conventional primary processing. Indeed, the business had recently purchased slaughtering equipment from a conventional grower and, once again, like their conventional counterparts, had a highly integrated production system in order to retain as much added value as possible. The use of chilling and freezing is at a similar level in both supply chains. Innovations present in conventional food supply chains such as refrigeration, packaging and cooking also apply to organic foods. Perhaps one of the best indications of the similarities between the conventional and organic chains is that conventional producers will now often have organic sidelines.

Organic potatoes are grown on a much smaller scale than conventional potatoes. Organic growers use different potato varieties than those popular with conventional growers owing to the use of different techniques such as the limited use of fertilisers and pesticides, and the necessity to rely on natural resistance of plants to disease and pests.

The retail sale of fresh organic potatoes is now well established, with supermarkets occupying the largest share of the market. This is a mixed blessing for organic growers. On the one side it provides access to large numbers of potential consumers. On the other it means that produce must meet the exacting standards of the supermarkets. Supermarkets pay high attention to the cosmetic appearance of foods, which greatly affects the competitiveness of organic producers. This is because organic potatoes may have blemishes in their appearance when compared to conventional produce, which affects the presentation that supermarkets value.

The industrial processing of organic potatoes and the manufacture of organic ready meals have been slow to take off, and therefore organic growers who produce crops below standards do not, unlike conventional producers, have alternative outlets. This leads to large amounts of wastage in organic potato production. According to an interview with an organic potato grower in the UK, up to 50 per cent of harvested organic potatoes can be considered as waste, as they are rejected by packers on the grounds of quality, size, colour and disease.

Rather like organic chicken production there are signs of absorption of the organic potato supply chain within the conventional model. For example, large seed producers, who deal with conventional seed, now also deal with organic seed. Similarly, some potato packers, who primarily deal with conventional potatoes, have also moved into the organic potato market and now also pack organic potatoes for their customers. Some potato processors, who process conventional potatoes, are looking into producing products from organic potatoes. Therefore, the functioning of the organic potato supply chain is highly dependent upon the operators of the conventional potato supply chain.

In terms of the structure of the supply chain, analysis indicates that from being marginal supply chains, organic food production and distribution has now been adopted by large mainstream producers, processors and retailers as part of their diversification strategy. It has become, essentially, a 'branded' form of production, which we may speculate offers all parties in the chain an opportunity to extract higher profits. Manufacturers are responding in kind to promote their brands in new markets. As Howard (2003, p. 3) reports:

> Many organic brands have been acquired by giant food processors such as General Mills, Kraft (Philip Morris) and Kellogg ... Slightly smaller global food processors are also establishing their own organic product lines (such as Dole, Chiquita and McCormick & Co) or acquiring organic brands.

In short, links between organic and conventional foods are now well established.

The UK experience of the interaction between organic and conventional supply chains is by no means unique:

> [I]n California agribusiness involvement does more than create a soft path of sustainability – an 'organic lite' if you will. For the conditions it sets undermine the ability of even the most committed producers to practice a purely alternative form of organic farming.
>
> (Guthman, 2004, pp. 301–302)

Guthman (2004, p. 307) goes on to suggest that agribusiness threatens organic farming in three ways. First, there is a political threat that leads to a lowering of organic standards. Second, there is the direct economic threat from agribusiness that reduces economic returns for organic farmers. Third, agribusiness may practice organic farming in a more shallow fashion that reduces the distinctiveness of organic production. To Guthman's list we might add a fourth point, that the close supply chain relationship between major conventional growers and the retailers allows the latter to encourage the former to enter organic markets. Supermarket chains now routinely sell organic foods alongside their conventional products. Sales of UK organic food are dominated by the supermarkets, accounting for about 71 per cent of sales (Soil Association, 2012). For some products, such as chicken, the supermarket share of organic sales may be even higher because conventional chicken is also widely distributed in the burgeoning food service industry (e.g. takeaways, restaurants and snack bars).

The food system and environmental impacts

One of the most comprehensive attempts to determine what evidence is available relating to the environmental impacts that occur in the life cycles of a range of food products was a Defra-funded project (Foster *et al.*, 2006), the so-called 'shopping trolley' study. The range of foods included in the study included both fresh and processed goods, organic and conventionally grown produce, locally sourced and globally sourced foods as well as taking account of different sources of nutrition. The sample used is a trolley of food types representative of the 150 highest-selling food items provided by a large retailer.

The review of evidence focused on studies that use the life cycle assessment (LCA) technique or a closely related approach. LCA considers the environmental impacts arising from the production, use and disposal of products, linking these to flows of substances between this system and the environment. LCA provides a mechanism for investigating and evaluating such impacts from the extraction of basic materials from nature, through material and component production, assembly, distribution, product use and end-of-life management (which may be disposal, reuse, recycling or recovery). The LCA methodologies also consider impacts on environmental media including air, water and land.

Overall, the review found that environmental impacts arising across the entire life cycle (including consumer activities and waste disposal) had been studied in detail for very few basic foods and even fewer processed foods. The bulk of the research that had been carried out focused on primary production, and only occasionally extended to cover processing. There are few studies taking account of the specific food system within the UK. Despite all the deficiencies in the data and the qualifications that are needed when applying it to specific foods and food types in the UK, the authors provide some general conclusions:

1 Organic vs conventionally grown foods: For many foods, the environmental impacts of organic agriculture are lower than for the equivalent conventionally grown food. However, it is not true for all foods and appears seldom to be true for all classes of environmental impact. In particular, organic agriculture can pose its own environmental problems in the production of some foods, either in terms of nutrient release to water or in terms of climate-change burdens. In short, there is insufficient evidence available to state that organic agriculture overall would have less of an environmental impact than conventional agriculture.
2 Local trolley vs globally sourced trolley: Evidence for a lower environmental impact of local preference in food supply and consumption overall is weak. Since there is a wide variation in the agricultural impacts of food grown in different parts of the world (e.g. in the amounts of water consumed), global sourcing could be a better environmental option for particular foods.
3 Fresh vs cold vs preserved food trolleys: The energy consumption involved in refrigeration means that a cold trolley will have higher environmental impacts than a fresh one. However, complications in interpretation arise because of the need to preserve food, coupled with uncertainty about wastage. Looking to the future, though, the growth of refrigeration as the 'default' method of food preservation and storage throughout the production–consumption system is likely to lead to higher impacts from energy generation.
4 Significance of transport in the life cycle: While the data are not clear-cut, there are suggestions that the environmental impacts of car-based shopping (and subsequent home cooking for some foods) are greater than those of transport within the distribution system itself. The environmental impacts of aviation are important for air-freighted products, but such products are a very small proportion of food consumed. However, with the volume of air-freighting of food items set to grow fast, aviation-related transport emissions are likely to become more significant in the future.

5 Significance of packaging: The environmental impact of packaging is certainly high for some foods (such as bottled drinks). However, quantifying the overall environmental impact of packaging involves assumptions about local practice regarding packaging waste (discard rates by consumers, predominance of different recovery or recycling mechanisms, etc.) and evidence of clear relevance to the UK is either sparse or inconclusive.

Conclusions: policy and research implications

The food crisis of 2007 and 2008 showed keenly the food system's exposure to the forces of globalisation. While increased prices had a devastating effect on developing countries, the price shocks also affected developed countries. In the UK, the rise in prices flowed through to the retail shelf; food price inflation peaked at 12.8 per cent in August 2008 (Rural and Environment Analytical Services, 2009). Consumers in the UK experienced a sudden reversal in a 26-year trend of year-on-year price reductions, challenging expectations of sources of ever-cheaper food. Retailers, suppliers and producers alike were caught by surprise by the sudden price rises. The events triggered widespread concerns over the global interdependency of modern food supply chains and demonstrated the political and social importance of affordable food. The Chatham House report (Ambler-Edwards *et al.*, 2009), among others, warned of the risk that pressures on global agriculture would herald a generational change in the experience of food supply; characterised by increased volatility and prices. Today, it could be argued that the continued trend of higher rates of food inflation is a reflection of global concerns as to the ability of agriculture to sustain growth and meet food security needs (see Morley *et al.*, Chapter 2, this volume).

The notion of developing new food systems is, of course, highly problematic. By adopting a system perspective that links together production and consumption we can begin to appreciate that changes on the farm, in the processing and manufacturing arenas must be accompanied by changes in consumption practices. Innovation in the food industry, for example, is linked to greater consumption of processed foods. Western consumers, particularly those in the UK, have very high ownership rates of freezers, fridges and microwave ovens that bind them to particular cooking practices. This has changed the way products are treated within the processing industry. For example, today, potatoes and chickens are often ingredients within complex ready meals, while before they were predominantly a major component in their own right of a household meal. Production of ready meals is an assembly process, where parts are sourced from various suppliers and competition over price drives manufacturers to source from abroad, thus changing further the economic and spatial geography of the food production systems.

From an environmental perspective there are several major areas, such as waste, energy, resource use and air pollution, where the food supply chain is having major impacts. Important technologies upon which the functioning of the modern food supply chain is based include refrigeration, transport and packaging. Penetration of these technologies at every stage of the supply chain and their wide use leads to high consumption of fuels, electric energy and materials, and leads to large amounts of waste and high volumes of greenhouse emissions. The food supply chain is very wasteful, not only with losses of harvests and food along the supply chain especially within the process and consumption stages, but also with increased use of packaging.

Since sustainability is a contested concept there is no single definition of a sustainable food system. Rather there are competing interpretations of food sustainability (e.g. organic foods, foods grown for local or regional markets). Alongside alternative versions of sustainability, individual foods have their own internal dynamic and potential for transition. We thus have to work with complex and multilayered notions of food systems and sustainability.

In these debates on possible sustainable futures for the food system, we can ask what is the role of the state? Where we may have looked for government to provide vision at a policy level and steering through policy delivery, as this volume illustrates, it is private sector interests and civil society who are now taking the lead in many areas. The withdrawal of the state from food markets and regulatory shortcomings mean that it has more limited capacity and legitimacy in the food system. Traditional notions of the state protecting and promoting the public interest – in this case in food sustainability – are weakened and the public as a consumer becomes dominant.

Inevitably, therefore, the more powerful interests in the food system will seek to shape both the debates and practices of sustainability. Private sector led interests are likely to promote narrow economic dominated notions of sustainability in which problems and solutions are constructed around ideas of flexibility of supply chains, global markets, efficiency in supply chains, reduced waste, technological innovations along the supply chain and consumer choice. Private sector dominated constructions of food sustainability will produce more efficient use of resources (e.g. land, materials and energy) and so will have benefits for these narrow notions of food sustainability, but change will take place within the system and remain 'managed' by key private sector actors in the food system.

External shocks, such as energy crises, may provoke compelling alternative narratives and practices on food system sustainability. It is equally possible, though, that tensions within the system become ever more difficult to contain. These tensions could erupt at any time, disrupting parts of or the whole of the system and encourage stronger forms of sustainability to emerge. For instance, the food system is particularly vulnerable to climate change. Changes in climate and water availability (e.g. droughts, flooding) have almost immediate effects upon crops and livestock. So, one key tension is: what happens when an innovative and adaptive food system confronts ever more challenging climatic conditions? One answer, of course is that as a new geography of production emerges, supply chains will uncouple and recouple, but will do so in new conditions of risk and uncertainty. Another tension occurs when resource efficiency runs up against resource limits. In both cases, business-as-usual with its limited view of sustainable food system becomes more untenable. Dislocation and disruption in parts of the food system create spaces for more inclusive and radical versions of food sustainability to emerge. These would have at their heart social and environmental concerns, as well as economic ones. This would place resource limits, environmental constraints and social inequality at the core of debates on food sustainability. While the conventional food system can promote innovations along and within supply chains, a stronger version of sustainability would challenge the organisation and values of the current system; it would involve a system change from production to consumption.

References

Ambler-Edwards, S., Bailey, K., Kiff, A., Lang, T., Lee, R., Marsden, T., Simons, D. and Tibbs, H. (2009) *Food Futures: Rethinking UK strategy*, London: Chatham House.

Bellows, A.C., Hamm, M.W. (2001) 'Local autonomy and sustainable development: Testing import substitution in localizing food systems', *Agricultural and Human Values*, vol. 18, pp. 271–284.

Bloom, J.D. and Hinrichs, C.C. (2010) 'Moving local food through conventional food system infrastructure: Value chain framework comparisons and insights', *Renewable Agriculture and Food Systems*, vol. 26, no. 1, pp. 13–23.

Competition Commission (2000) *Supermarkets: A report on the supply of groceries from multiple stores in the United Kingdom*, London: Competition Commission.

Competition Commission (2009) *The Groceries (Supply Chain Practices) Market Investigation Order 2009*, London: Competition Commission.

Coombs, R., Harvey, M. and Tether, B. (2003) 'Analysing distributed processes of provision and innovation', *Journal of Corporate Change*, vol. 12, no. 6, pp. 1125–1155.

Cox, A. and Chicksand, D. (2005) 'The limits of lean management thinking: Multiple retailers and food and farming supply chains', *European Management Journal*, vol. 23, no. 6, pp. 648–662.

Cox, A., Chicksand, D. and Palmer, M. (2007) 'Stairways to heaven or treadmills to oblivion? Creating sustainable strategies in red meat supply chains', *British Food Journal*, vol. 109, no. 9, pp. 689–720.

Curran, M.A. (1996) *Environmental Life Cycle Assessment*, New York: McGraw-Hill.

Defra (2009) *Agriculture in the UK*, London: Defra.

Edwards-Jones, G., Milà i Canals, L., Hounsome, N., Truninger, M., Koerber, G., Hounsome, B., Cross, P., York, E.H., Hospido, A., Plassmann, K., Harris, I.M., Edwards, R.T., Day, G.A.S., Tomos, A.D., Cowell, S.J. and Jones, D.L. (2008) 'Testing the assertion that "local food is best": The challenges of an evidence-based approach', *Trends in Food Science and Technology*, vol. 19, no. 5, pp. 265–274.

Feindt, P. and Flynn, A. (2009) 'Policy stretching and institutional layering: British food policy between security, safety, quality, health and climate change', *British Politics*, vol. 4, no. 3, pp. 386–414.

Food Futures Partnership (2007) *A Food Strategy for Manchester 2007*, Manchester: Manchester Joint Health Unit.

Foresight (2011) *The Future of Food and Farming, Final Project Report*, London: The Government Office for Science.

Foster, C., Green, K., Bleda, M., Dewick, P., Flynn, A. and Mylan, J. (2006) *Environmental Impacts of Food Production and Consumption: A report to the Department for the Environment, Food and Rural Affairs, Manchester Business School*, London: Defra.

Francis, M., Simons, D. and Bourlakis, M. (2008) 'Value chain analysis in the UK beef foodservice sector', *Supply Chain Management*, vol. 13, no. 1, pp. 83–91.

Fuchs, D., Kalfagianni, A. and Havinga, T. (2011) 'Actors in private food governance: The legitimacy of retail standards and multistakeholder initiatives with civil society participation', *Agriculture and Human Values*, vol. 28, no. 3, pp. 353–367.

Fulponi, L. (2006) 'Private voluntary standards in the food system: The perspective of major food retailers in OECD countries', *Food Policy*, vol. 31, no. 1, pp. 1–13.

Garnett, T. (2003) *Wise Moves: Exploring the relationships between food, transport and carbon dioxide*. London: Transport 2000 Trust.

Garnett, T. and Godfray, C. (2012) *Sustainable Intensification in Agriculture: Navigating a course through competing food system priorities*, Food Climate Research Network and the Oxford Martin Programme on the Future of Food, Oxford: University of Oxford.

Geels, F.W. and Schot, J. (2007) 'Typology of sociotechnical transition pathways', *Research Policy*, vol. 36, no. 3, pp. 399–417.

Godfray, H.C.J., Crute, I.R., Haddad, L., Lawrence, D., Muir, J.F., Nisbett, N., Pretty, J., Robinson, S., Toulmin, C. and Whiteley, R. (2010) 'The future of the global food system', *Philosophical Transaction of the Royal Society B*, vol. 365, pp. 2769–2777.

Green, K., Harvey, M. and Mcmeekin, A. (2003) 'Transformations in food consumption and production systems', *Journal of Environmental Policy and Planning*, vol. 5, no. 2, pp. 145–163.

Guptill, A. and Wilkins, J.L. (2002) 'Buying into the food system: Trends in food retailing in the US and implications for local foods', *Agriculture and Human Values*, vol. 19, no. 1, pp. 39–51.

Guthman, J. (2004) 'The trouble with "Organic Lite" in California: A rejoinder to the "conventionalisation" debate', *Sociologia Ruralis*, vol. 44, no. 3, pp. 301–316.

Hall, J. (2000) 'Environmental supply chain dynamics', *Journal of Cleaner Production*, vol. 8, no. 6, pp. 455–471.

Havinga, T. (2012) *Transitions in Food Governance in Europe*, Nijmegen Sociology of Law Working Papers Series 2012/02.

Hendrickson, M.K. and Heffernan, W.D. (2002) 'Opening spaces through relocalization: Locating potential resistance in the weaknesses of the global food system', *Sociologia Ruralis*, vol. 42, no. 4, pp. 347–369.

Hendrickson, M.K., Heffernan, W.D., Howard, P.H. and Heffernan, J.D. (2001) *Consolidation in Food Retailing and Dairy: Implications for farmers and consumers in a global food system*, National Farmers Union.

Hingley, M.K. (2005) 'Power imbalance in UK agri-food supply channels: Learning to live with the supermarkets?', *Journal of Marketing Management*, vol. 21, no. 1–2, pp. 63–68.

House of Commons Environment, Food and Rural Affairs Committee (2009) *Securing Food Supplies up to 2050: The challenges faced by the UK*, Fourth Report of Session 2008–09, Volume I, HC 213-I, London: The Stationery Office.

Howard, P. (2003) 'Consolidation in food and agriculture', *CCOF Magazine*, Winter 2003–2004, pp. 2–6.

IAASTD (2009) *Agriculture at a Crossroads*, Global Report, International Assessment of Agricultural Knowledge, Science and Technology for Development, Washington, DC: Island Press.

Ingram, J., Ericksen, P. and Liverman, D. (eds) (2010) *Food Security and Global Environmental Change*, London: Earthscan.

Ireland, R. and Bruce, R. (2000) CPFR: Only the beginning of collaboration, *Supply Chain Management Review*, September/October, pp. 80–88.

Jungbluth, N., Tietje, O. and Scholz, R.W. (2000) 'Food purchases: Impacts from the consumers' point of view investigated with a modular LCA', *International Journal of Life Cycle Assessment*, vol. 5, no. 3, pp. 134–142.

Kirwan, J. (2004) 'Alternative strategies in the UK agro-food system: Interrogating the alterity of farmers' markets', *Sociologia Ruralis*, vol. 44, no. 4, pp. 395–415.

Konefal, J., Mascarenhas, M. and Hatanaka, M. (2005) 'Governance in the global agro-food system: Backlighting the role of transnational supermarket chains', *Agriculture and Human Values*, vol. 22, no. 3, pp. 291–302.

Marks & Spencer (2010) *Our Plan A Commitments 2010–2015*. London: Marks & Spencer Plc.

Marks & Spencer (2012) *How we do Business Report*. London: Marks & Spencer Plc.

Marsden, T. (2004) 'Theorising food quality: Some key issues in understanding its competitive production and regulation', in M. Harvey, A. McMeekin and A. Warde (eds) *Qualities of Food*, Manchester: Manchester University Press, pp. 129–155.

Marsden, T., Flynn, A. and Harrison, M. (2000) *Consuming Interests: The social provision of foods*, London: UCL Press.

Marsden, T., Flynn, A. and Ward, N. (1994) 'Food regulation in Britain: A national system in an international context', in A. Bonanno, L. Busch, W. Friedland, L. Gouveia and E. Mingione (eds) *From Columbus to ConAgra: The globalization of agriculture and food*, Kansas City: University Press of Kansas, pp. 105–124.

Matos, S. and Hall, J. (2007) 'Integrating sustainable development in the supply chain: The case of life cycle assessment in oil and gas and agricultural biotechnology', *Journal of Operations Management*, vol. 25, no. 6, pp. 1083–1102.

Mintel (2012) *Private Label Food and Drink in the UK*, Mintel Oxygen Report, March.

Morris, C. and Kirwan, J. (2011) 'Exploring the ecological dimensions of producer strategies in alternative food networks in the UK', *Sociologia Ruralis*, vol. 5, no. 4, pp. 349–369.

Morrisons (2010) *Corporate Social Responsibility Report 2009/2010*, Bradford: Morrisons Supermarkets Plc.

Morrisons (2012) *Corporate Responsibility Review 2011/12. Food with thought*, Bradford: Morrisons Supermarkets Plc.

Murdoch, J. and Meile, M. (2004) 'A new aesthetic of food? Relational reflexivity in the "alternative" food movement', in M. Harvey, A. McMeekin and A. Warde (eds) *Qualities of Food*, Manchester: Manchester University Press, pp. 156–175.

Oosterveer, P. (2012) 'Restructuring food supply: Sustainability and supermarkets', in G. Spaargarren, P. Oosterveer and A. Loeber (eds) *Food Practices in Transition: Changing food consumption, retail and production in the age of reflexive modernity*, Abingdon: Routledge, pp. 153–176.

Pretty, J. (2008) 'Agricultural sustainability: Concepts, principles and evidence', *Philosophical Transactions of the Royal Society B: Biological Sciences*, vol. 363, no. 1491, pp. 447–465.

Ring, P. and Van de Ven, A. (1992) 'Structuring co-operative relationships between organizations', *Strategic Management Review*, vol. 13, pp. 483–498.

Rural and Environment Analytical Services (2009) *Food Prices: An overview of current evidence*, Rural and Environment Research and Analysis Directorate: The Scottish Government.

Sainsbury's (2012) *20 x 20 Our 20 Commitments to Help Us All Live Well For Less: Our progress so far*, London: J Sainsbury Plc.

Smith, B.G. (2008) 'Developing sustainable food supply chains', *Philosophical Transactions of the Royal Society B: Biological Sciences*, vol. 363, no. 1492, pp. 849–861.

Soil Association (2012) *Organic Market Report 2012*, Bristol: Soil Association.

Sonnino, R. and Marsden, T. (2006) 'Beyond the divide: Rethinking relationships between alternative and conventional food networks in Europe', *Journal of Economic Geography*, vol. 6, no. 2, pp. 181–199.

Southerton, D., Chappelles, H. and van Vliet, B. (2004) *Sustainable Consumption: The implications of changing infrastructure*, Cheltenham: Edward Elgar.

Spence, L. and Bourlakis, M. (2009) 'The evolution from corporate social responsibility to supply chain responsibility: The case of Waitrose', *Supply Chain Management: An International Journal*, vol. 14, no. 4, pp. 291–302.

Storey, J., Emberson, C., Godsell, J. and Harrison, A. (2006) 'Supply chain management: Theory, practice and future challenges', *International Journal of Operations & Production Management*, vol. 26, no. 7, pp. 754–774.

Sustainable Development Commission (SDC) (2009) *Food Security and Sustainability*, SDC position paper: Sustainable Development Commission.

Tesco (2012) 'It's at the heart of what we do, Tesco Corporate Responsibility Review', Cheshunt: Tesco Plc.

The Royal Society (2009) *Reaping the Benefits: Science and the sustainable intensification of global agriculture*, London: The Royal Society.

Tregear, A. (2011) 'Progressing knowledge in alternative and local food networks: Critical reflections and a research agenda', *Journal of Rural Studies*, vol. 27, no. 4, pp. 419–430.

Wall Street Journal Europe Edition (2012) 'Conagra shows faith in private-label foods', 27 November.

6 Biosecurity and the bioeconomy

The case of disease regulation in the UK and New Zealand

Gareth Enticott

Introduction

Biosecurity has emerged as a global object of policy to facilitate and ensure the continuation of the bioeconomy – the free and safe movement of agricultural animals, products and practices across the globe. Of key concern for biosecurity has been the ability to control animal disease and disease vectors in order to encourage the most effective production of food. Here there are links with other global security agendas: there are strong connections between initiatives to promote biosecurity and food security (Waage and Mumford, 2008), while the Food and Agriculture Organization's (FAO) 'One Health' campaign stresses the interconnections between human and animal health.

Yet, despite biosecurity's apparent centrality to life, its significance as an object of policy owes much to a changing international and national landscape of agricultural governance. In many ways, these changes, made in association with the pursuit of the neoliberal trade liberalization agenda, have in effect given life to biosecurity. Indeed failures within the bioeconomy – the outbreak of Foot and Mouth Disease (FMD) in 2001, the threat of new zoonotic diseases such as pandemic avian influenza (H5N1 or bird flu), and the spread of existing diseases such as bluetongue virus to areas that were once free of disease – have brought the very idea of biosecurity to the public's attention. Before 2001 you would have struggled to find any mention of biosecurity in a national newspaper in Great Britain. By 2010, however, biosecurity had become connected to a broader agricultural securitization agenda: the spread of animal disease, climate change and energy crises that are part of the 'perfect storm' that could lead to worldwide food shortages (Ilbery, 2012). In this climate of fear and catastrophe, biosecurity has become a solution to the 'unknown unknowns' of the bioeconomy (Braun, 2007).

But these uses of biosecurity also hint at something else: that despite (or perhaps because of them – see Law (2006)) international laws seeking to standardize biosecurity practices, the assembling and practice of biosecurity remains uneven, influenced by local affairs and practices, which can result in significant policy variations. In practice, the universalizing tendencies of neoliberalism and the bioeconomy highlight instead the difficulty of transferring policies between contexts, and point to the productive effort of agricultural actors and relationships across local and international, and public and private spaces required to make biosecurity initiatives work in various ways.

The aim of this chapter is therefore to explore the evolution of biosecurity governance and its relationship with the agricultural bioeconomy: how biosecurity is made to work, how it varies across spatial scales, and how local social interactions result in different translations of biosecurity policies. To do this, the chapter presents a comparative analysis of a specific cattle disease – bovine Tuberculosis – that has had a significant impact on the agricultural industry in New Zealand and Great Britain. The analysis reveals how biosecurity policies rely on different scalar constructions of biosecurity, and that while biosecurity practices can be as mobile as the disease itself, they must nevertheless be worked out from struggles between public and private actors to make biosecurity work across different local contexts. In conclusion, the chapter considers whether these neoliberal biosecurities offer the best form of resilience for the future of agriculture.

The spatial scales of biosecurity

As a way of describing the emergence of biosecurity, its relationship to the bioeconomy and its effects on the geography of agriculture, this section presents three ways in which biosecurity is practised and understood. Each of these versions constructs biosecurity at a particular spatial scale. The first version presents biosecurity as a local solution to local agricultural problems; the second as individual responsibility towards national space; and the third as a form of territorial responsibility, performed across international space (see Figure 6.1).

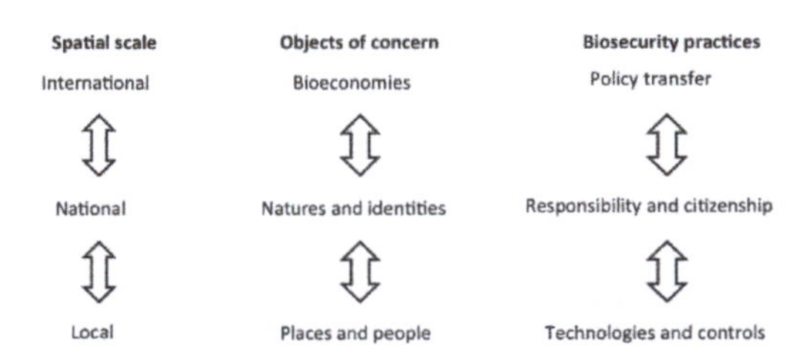

Figure 6.1 Biosecurity objects and practices.

Disaster biosecurity: the local scale

The idea of biosecurity has been shaped by a series of recent agricultural disasters, without which there would be little consciousness of biosecurity among the public or policy makers (Donaldson, 2008). These disasters have done much to cement the idea that biosecurity is a solution to a set of social and economic problems caused by animal disease, but experienced at the local scale – that is, on farms and in villages in rural areas.

The newness and localness of biosecurity is apparent when considering the impacts of outbreaks of FMD and other agricultural diseases. It was not until 1996 that biosecurity first appeared in a UK newspaper in an article warning of the dangers posed by introduced species to the natural environment (Pearce, 1996). Since then, however, biosecurity has come to be framed in different ways and linked to agriculture and the exclusion and eradication of animal disease. As Figure 6.2 shows, the handful of mentions of biosecurity before 2000 were dwarfed by those in 2001 and 2007 – both years in which outbreaks of FMD ravaged agriculture across the UK.

Thanks to the veterinary sciences, a number of different ways of preventing and combatting animal disease have been identified. These range from animal husbandry methods and herd management systems, through to genetic manipulation of livestock and the development of vaccines. Seen in this light, it is a wonder that biosecurity has appeared so recently as a concept in animal disease prevention. It is, after all, something that forms the foundation of agriculture and the veterinary professions: who doesn't want to be healthy, eat food from healthy animals, and live in healthy surroundings? But at the same time, biosecurity has become associated with particular activities involving hygiene and cleanliness. The image of disinfecting the wheels of lorries and other agricultural vehicles is one to which biosecurity is frequently reduced. Yet, it is also something of a symbolic solution, more a way of coping with uncertainty, than of eliminating it; or a means to blame others, or oneself, for disease outbreaks (Nerlich and Wright, 2006).

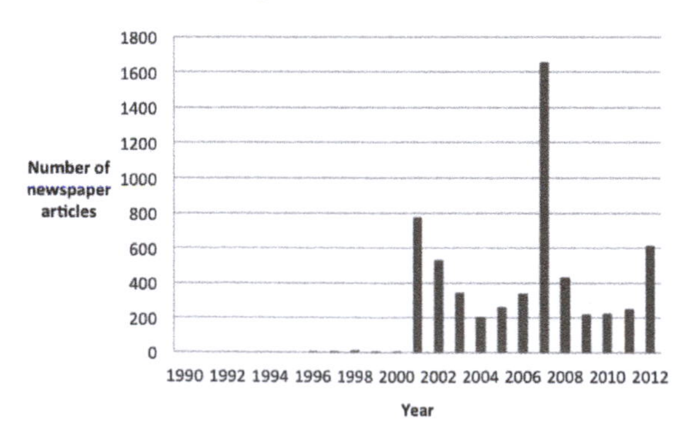

Figure 6.2 Number of UK newspaper articles mentioning biosecurity.

Indeed, while biosecurity failures can result in agricultural crises, their effects are as much human as they are animal. In the aftermath of the FMD crisis, studies revealed the economic impacts not just to farm businesses, but also other rural enterprises (Bennett and Phillipson, 2004). Farmers' well-being and mental health was severely dented by the loss of animals as well as the isolation and break in agricultural and social routines (Mort *et al.*, 2005; Convery *et al.*, 2008). The outbreak also affected the institutional structures of disease control and the practices of disease control planning (Ward *et al.*, 2004; Wilkinson, 2011). Calls to cull badgers to help reduce levels of bovine Tuberculosis in cattle have also become increasingly framed as a solution to farmers' stress and mental health problems. In these ways then, biosecurity reflects a set of techniques to deal with the locally experienced effects of animal disease.

Biosecurity citizenship at the national scale

While biosecurity can help alleviate the effects of diseases at a local level, it is also associated with the development of biosecurity subjectivities that seek to develop responsible behaviour across the nation-state. This duty of care towards agriculture at a national level is not immediately obvious from models used to understand the phases of biological and disease invasion. Frequently, biosecurity breaches are imagined in linear stages, involving distinct chronological and biological pathways. Williamson (2006) and Lockwood *et al.* (2005) describe a similar staged process where biosecurity hazards are first imported into a country, released or escape into the wild, become established, spread to other locations/populations and become a problem that people want to deal with. Others characterize these stages in terms of human activity, such as prevention, anticipation, management, alleviation and adaptation (Fish *et al.*, 2011).

However, these geometric retellings of biosecurity incursions tend to underplay problems internal to biosecurity practices, viewing biosecurity threats as external rather than lying within its very system (Law, 2006; Hinchliffe *et al.*, 2012). While this suggests the need to build resilient agro-ecological systems (Waage and Mumford, 2008), it also highlights how duties of care and responsibility can change the course of biosecurity events.

Developing a responsibility or duty to maintain good biosecurity for the benefit of the national herd has become a central tenet in animal disease policy. In the UK, the FMD crisis alerted government to the financial cost of state-led eradication programmes. As a result, the development of biosecurity responsibility has become tied up in neoliberal programmes of state reform in which disease control costs are shifted from the state to the agricultural sector. While this can take many forms, Barker (2008, 2010) shows how it can result in the development of new biosecurity subjectivities that create a biosecure citizenship that stresses individual responsibility.

Barker (2010) suggests three ways in which biosecure citizenship is constructed. First, publics are encouraged by governments to think not just of governing their biological bodies, but also the eco-relational body. Techniques of persuasion encourage us to think about and manage these symbiotic interactions with the environment. Second, biosecure citizenship is formed out of an alignment of the symbiotic self with national identity. Drawing on New Zealand's Biosecurity Act, Barker shows how it enshrines a duty to care for national biological symbols (such as Kiwis and Kauri trees), classifies the spatial belonging/alienation of natures, and constructs spaces and practices designed to protect them. Third, there is a blurring of the boundaries of the public and private as public biosecurity duties are applied to private property but in non-contractual ways that are reliant on public willingness and the effectiveness of attempts to normalize pro-biosecurity citizenship. In these ways, biosecurity moves from the local to the national scale: it involves collectively securing objects of national importance, changing behaviour in the name of national identity, and reconfiguring national institutions to do so.

Territorial and international responsibilities

Biosecure citizenship highlights the importance of national identity and territory. But in a third version of biosecurity, responsibility stretches to maintaining an international market-based territory. Attempts to prevent the spread of disease while ensuring free trade form the basis to this international geography beginning in 1924, when the Office International des Epizooties (OIE, also known as the World Organisation for Animal Health) was formed. The OIE's Terrestrial Animal Health Code sets standards and protocols for the diagnosis of disease and the criteria for a country to be officially recognized as disease-free. The 172 worldwide members of the OIE are obliged to report incidents of those diseases listed as notifiable. Moreover, the OIE acts as a reference body for the World Trade Organization whose Sanitary and Phytosanitary (SPS) agreement enforces the codes set out by the OIE to protect animal health and ensure free trade. For the UK, these codes are also incorporated into European legislation governing the trade of animals and animal products between member states.

For Braun (2007), these activities reveal the connections between biopolitics and geopolitics. Biosecurity is not simply a local or national affair: biosecurity must be taken to distant territories in order to protect those closer to home. Thus, 'biosecurity names an answer to the problem of the mutability and unpredictability of biological life within a political–economic order that is premised upon global economic integration' (Braun, 2007, p. 19). So, while territory and national identity are central to Barker's description of biosecurity subjectivity, it is also true that biosecurity solutions are dependent on acts of deterritorialization. Moreover, while biosecurity has come to justify 'the global extension of forms of sovereign power' (Braun, 2007,

p. 6), biosecurity imagines certain biological futures in favour of others. This may lead to the displacing of 'other' explanations, knowledges and practices of animal disease (Enticott and Wilkinson, 2013), the valuing of certain diseases over others, and the global movement of solutions and practices to resolve the challenges of animal disease. Biosecurity is more than a set of benign practices: it is a global project in the name of a particular community, which reconfigures the relationships between nature and society in other communities (Braun, 2007).

This de/reterritorialization of agricultural space is dependent on the tools of neoliberalism. The standardization of disease protocols and techniques of risk assessment have become key biosecurity technologies; simultaneously preventing the spread of disease while maintaining free trade, much as they have been throughout agriculture (Busch, 2010). These techniques can create new institutional arrangements by deregulating, shifting regulation from the public to the private sector, or creating new hybrid partnerships and organizations to secure access to the global spaces of agriculture.

The techniques associated with the international management of disease are also mobile. Like other forms of neoliberal policy (see Peck, 2002), so can specific tools and practices be transferred from one country to another, particularly where one country has successfully eradicated a disease. Yet this is not as simple as it sounds: practices may require translating or reassembling to fit local circumstances (Prince, 2010), highlighting how neoliberalism itself is something of a 'rascal concept' from which 'messy hybrids' unfold in an uneven manner (Peck, 2010; Brenner *et al.*, 2010).

In this respect, the neoliberal transformations of agricultural and biosecurity institutions may not simply be read off. Rather, as Hodge and Adams (Hodge and Adams, 2012) put it, neoliberalism is a form of 'institutional blending': solutions are worked out unevenly in different contexts by a wide range of actors through a complex mixing of processes. For biosecurity, it is not pre-given that the international spaces of biosecurity will be implemented or implemented in the same way in all places (Dibden *et al.*, 2011). Importantly, it is the tensions between the national and international spaces of biosecurity that give impetus to these differing versions of biosecurity. Thus, biosecurity standards are given a local character as they emerge 'out of complex articulations between actors in multiple locations' (Higgins and Larner, 2010, p. 10) to make them workable across agricultural space. Despite attempts to harmonize trade rules and create international biosecure territories, perceptions of the risks to national biosecurity identities can still lead to a divergence in biosecurity practices and a protection of national agricultural territory (Higgins and Dibden, 2011; Higgins *et al.*, 2012; Maye *et al.*, 2012).

Although biosecurity is put to use at different spatial scales, interactions and dependencies between these different forms of biosecurity are likely to affect the extent to which biosecurity solutions work and/

or vary between places. For instance, the implementation of veterinary practices at a local level may depend on cultural attitudes towards disease and nature, and their complementarity may (or may not) affect the way international agreements are considered. Similarly, the international transfer of policy approaches to disease eradication may result in various translations of these policies, depending on the way they interact with local and national enactments of biosecurity. Thus, tracing these interactions and translations can help reveal how the agricultural bioeconomy is enacted in various and uneven ways, involving different processes and institutional hybrids. For the agricultural bioeconomy, biosecurity policies may emphasize different versions of biosecure territory, enact and protect different versions of nature for the benefit of different agricultural communities.

Neoliberalism and animal disease: the case of bovine Tuberculosis

Bovine Tuberculosis (bTB) is a good case to explore the different unfoldings of neoliberal biosecurity. The connections between the bioeconomy and the management of bTB reach back to when the disease was recognized as a human and animal problem in the late nineteenth century. While farmers were concerned about economic losses from infected cattle, public health professionals were concerned about the thousands of people dying from the consumption of infected beef and milk (Waddington, 2004). Given these human and animal health impacts, bTB was therefore an early candidate for creating and standardizing a set of international controls under the OIE's animal health code. In short, for over 100 years, activities aimed at preventing bTB have been embedded within systems of national and international agricultural governance.

Two countries that have been severely affected by bTB have been Great Britain and New Zealand. In Great Britain, high levels of bTB in the early part of the twentieth century were brought under control by a state-led national eradication scheme. By 1965, it was thought that the disease was almost eradicated (MAFF, 1965). The following decades, however, saw a gradual increase, which was attributed to badgers spreading the disease to cattle. Despite contentious badger removal operations and scientific research, bTB has continued to rise such that by 2010 over 35,000 cattle were slaughtered as a result of the disease (see Figure 6.3). There are some similarities with the disease in New Zealand. Endemic levels of bTB were gradually reduced following a government-led eradication campaign beginning in the late 1950s. Like in the UK, these efforts were compromised by the discovery that wildlife – in this case brushtail possums – could spread the disease to cattle. However, unlike the UK, the disease has declined, such that it is now rare in many parts of the country (see Figure 6.3).

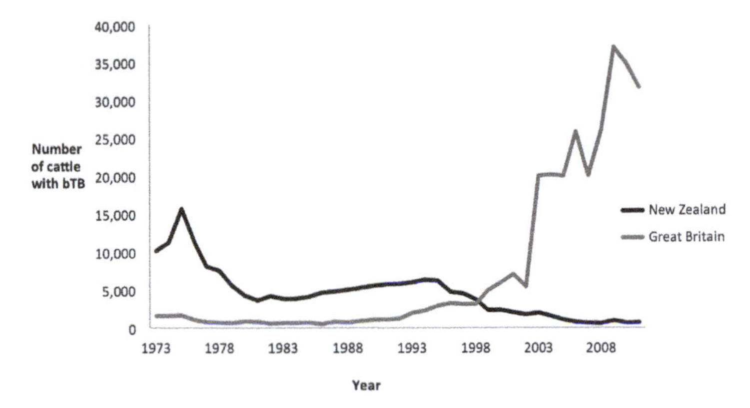

Figure 6.3 Bovine Tuberculosis in Great Britain and New Zealand.

Another difference has been in the styles of government used to eradicate the disease. In both Great Britain and New Zealand, various forms of liberal and neoliberal government have been deployed in attempts to manage the disease, but there have also been attempts to transfer ideas from New Zealand into the management of bTB in Great Britain. The following sections document the impact of these neoliberal forms of animal health governance, comparing how different versions of neoliberalism and disease are enacted, and the various ways through which the agricultural industry has become involved in disease management.

New Zealand

In New Zealand, the eradication of bTB is managed by a farmer-led organization, known as the Animal Health Board (AHB). Farmers pay for and decide upon for themselves voluntary and/or mandatory control measures. Some aspects of the scheme – such as bTB testing – are delivered by private contractors through a system of competitive tendering. In all of this, the State lurks malignly in the background. If this is the ideal neoliberal model, then the scheme has to be understood in the context of much broader socio-economic changes in New Zealand and its historical approach to dealing with animal disease.

In contrast to the current position, the origins of New Zealand's bTB control programme were undeniably paternalistic. While the Crown (i.e. national government) had resisted involvement during the mid-twentieth century out of fear of the costs of any eradication scheme, by the late 1950s industry-led voluntary eradication programmes had been overtaken by a scheme run by the Department of Agriculture. At its heart, the scheme had the national economy as its object of concern. While public health had originally been a driver of bTB eradication, these concerns were quietly replaced by the desire to protect New Zealand's exports of beef and dairy products. It was on this basis that it 'became vital that New Zealand embark on a tuberculosis eradication scheme encompassing the entire national herd' (Davidson, 2002, p. 9).

The resulting scheme was created and overseen by Dr Sam Jamieson, a Scottish vet, who ran the scheme in an autocratic centralized fashion. Believing that bTB was out of control, Jamieson rebuffed concerns from vets and farmers about the effect a strict eradication policy would have on farmers' livelihoods, dismissing them as 'unscientific'. Nevertheless, Jamieson recognized that farmers needed to be included in aspects of decision making, arguing that only when 'all dairy farmers cooperate it will be possible to stamp out entirely a disease which is both unnecessary and harmful to the health of animals and human beings alike' (Jamieson, 1960, p. 8). This led to the creation of local advisory committees involving farmers, vets and Department of Agriculture officials in which farmers allocated 'hardship payments' to others disproportionately affected by bTB. In the 1970s the committees' remit was broadened – they became Regional Animal Health Advice Committees (RAHACs). Although the RAHACs were not meant to be involved in operational decisions, increasingly they were. Farmers, pressing for the use of new diagnostic tests found they were influential. The governance of animal disease appeared on the cusp of a new style of governing. It was a time when farmer ownership, involvement, and control were starting to grow and replace a paternalistic scheme that had been imposed by the Crown. By the mid 1990s, the system had moved to a model where bTB was seen as farmers' problem, while the Crown would support but not tell farmers what to do.

The shift was precipitated by two financial crises. The first was the withdrawal of government funding for bTB control in the late 1970s and the 1980s. The financial resources of disease control had been stretched as a wildlife vector – the brushtailed possum – was identified, requiring trapping and poisoning. By the late 1970s, the disease had been reduced significantly and with eradication looking likely, politicians whose priorities lay outside agriculture decided to cut funding. The effect was a marked increase in levels of bTB.

A second financial crisis came in 1984 when New Zealand's finance minister, Roger Douglas, began a series of financial reforms, beginning with the removal of subsidies from New Zealand's farmers. The Ministry of Agriculture, Fisheries and Food (MAFF) was downsized and externalized as a way of introducing competition to the services it provided. In 1987 the number of staff reporting to the Chief Veterinary Officer was over 2,000; by 1991 it was just 31. For the management of bTB, functions such as bTB testing – traditionally undertaken by an army of MAFF technicians – were shifted to a new state owned enterprise (SOE) called MAFF Quality Management (MAFFQual). For veterinarians working in MAFF, these changes affected their ability to maintain contact with field staff conducting bTB testing and plan what needed to be done about the disease. Later, MAFFQual was spilt into two organizations – Agriquality and Asure – and invited to tender for bTB testing, thereby introducing competition to bTB testing where once there had been none.

Simultaneously, the government had withdrawn finance, radically altered the organizational structure of bTB field and strategic operations, and withdrawn financial subsidies to farmers. But, from this chaos the farmer-led AHB emerged to replace the old regime. To a degree, the new structure already existed in the shape of the National Animal Health Advisory Committee (NAHAC), and their regional equivalents (the RAHACs). In 1988 the NAHAC officially became the Animal Health Board, a partnership between the agricultural industry and the government. Further legislation was required before the AHB could take control of bTB. In 1993, the Biosecurity Act was passed allowing the AHB to become the management agency for the bTB National Pest Management Strategy (NPMS). The rules and constitution of the AHB were certified in 1996 and in 1998, the AHB's NPMS was approved by MAFF, giving the AHB legal status to pursue its mission of eradicating bovine Tuberculosis from New Zealand.

Under the terms of the NPMS, the costs of disease control would be shared between 'beneficiaries' (farmers) and 'exacerbators' (owners of land harbouring possums – principally the Crown). In 2012, the majority (approximately 55 per cent) of the AHB's income was contributed by the agricultural industry; 35 per cent by the New Zealand government; and 6 per cent from the regional councils of New Zealand (also considered exacerbators) (Animal Health Board, 2012). The AHB raises funds from farmers using a levy, currently set at NZ$11.50 on all slaughtered adult cattle, and NZ$0.01 per kg of milk solids. Approximately 40 per cent of farmers' levies are spent on disease control activities (bTB testing, compensation and management) with the remainder directed to possum control. All of the State's contribution covers possum control activities. The establishment of the AHB as a pest management agency also conferred legal powers upon the AHB that allowed it to negotiate contracts with the private sector to deliver bTB testing. The introduction of competitive structures matched the prevailing managerial ideologies, but was also driven by a desire to reduce and control costs as far as possible – after all, the AHB was mostly industry-owned with a motto reading 'farmer pays, farmer says', requiring a different mindset among the vets and officials who had once worked for MAFF.

The approach that developed in New Zealand is interesting for a number of reasons. First, it highlights how the transition to a neoliberal mode of disease control was by no means easy, and fraught with contradiction. Although the new structure lay outside of government, for the early years at least, it remained reliant upon them. Traditional government institutions were required to actively shape new organizational structures after they had lost responsibility for them. The realization that the government were stepping back from funding bTB eradication alarmed vets in MAFF, but also meant that they would have to encourage farmers to take on the mantle if the disease was to be addressed. The Chief Veterinary Officer at the time was in no doubt that should happen, arguing that: 'MAFF could no longer be paternalistic – we had to be a partner, and essentially a partner that was withdrawing as well in terms of the decision making'. This, though, was easier said than done. The withdrawal of subsidies had left the industry shell-shocked, just at a time when it was being asked to take control. As a result, the new structure took time to take shape with MAF staff having to lead the AHB for its formative years.

Second, in this new era of disease control, farmers were also enrolled in a new set of biosecurity subjectivities (cf. Barker, 2010). These biosecure identities were not simply a function of their new financial commitment to the eradication of bTB, but through other non-statutory efforts to encourage farmers to act in more biosecure ways and contribute to the national effort of ridding New Zealand of bTB. An example of this was the development of a risk-based trading scheme to limit the movements of cattle between infected and clean areas of New Zealand. Farms were classified according to the number of years they were bTB-free (e.g. C1, C2 through to C10 where C stands for clear and the number the years since the last positive test). While this conferred status on certain herds, the scheme also penalized farmers purchasing cattle from herds with inferior bTB status: were a C10 farm to buy cattle from a C5 herd, it would adopt the lower status classification. Other attempts to normalize biosecure farming can be seen in the lower rate of compensation offered to farmers. Farmers felt that a compensation rate of 65 per cent of full market value would discourage the perverse incentives that compensation schemes can create. Importantly, both these measures to develop a biosecure citizenship were decisions taken by farmers.

Third, the roll-out of neoliberal solutions was dependent on context. The creation of the AHB had been facilitated by government-initiated rounds of de-professionalization within the veterinary service dating back to the 1950s. The demand for bTB testing required to eradicate bTB was never going to be met by New Zealand's small veterinary service. Enlisting the support of Federated Farmers – New Zealand's national farming union – Sam Jamieson successfully lobbied for a change in the law to allow trained technicians to conduct bTB tests. The veterinary profession argued that the move would financially detriment private practices, but the absence of sufficient vets and a long-standing tradition of using trained technicians meant the opposition received short shrift.

Cultural attitudes towards nature in New Zealand also provided an environment in which the control of wildlife could be handed over to a non-government organization. That possums were a non-native species – from Australia, no less – responsible for destroying native Kiwi was a fortunate coincidence: it mobilized a national symbol central to the New Zealand identity in a way that helped justify the extensive poisoning operations of possums to deal with bTB in cattle. The synergy between nature and agricultural protection has proved beneficial in supporting the AHBs possum control operations despite growing opposition to the use of poison to kill possums. Indeed, the eradication of bTB is increasingly justified in terms of its role in biodiversity protection, rather than the national economy (see, for example, PCE (2011)).

While these historical and cultural factors underpin the development of the New Zealand approach to bTB, they also highlight how dependent it was on a unique set of circumstances. Other countries may admire New Zealand's neoliberal approach, but the context of disease control limits the mobility of policy. As many farmers point out, the management of the disease would be wholly different if native species spread bTB. Given the set of circumstances in which bTB policy has evolved, it would require significant revision and adaptation.

Great Britain

Unlike New Zealand, Great Britain (with the exception of Scotland) is not close to eradicating bovine Tuberculosis. In fact, since 2001, the level of disease has been rising at alarming rates (see Figure 6.3) with the cost to the taxpayer approaching £100 million a year. Many farmers have lost faith in the government's desire and ability to do anything about the disease (Enticott, 2008). For others, the effects of living with bTB had driven them to despair, damaging the financial viability of their business and in some cases, driving them to illegal measures in a bid to safeguard their farms (Enticott, 2011).

Delays and confusion over what action to take have played a fundamental role in the increase of bTB. As Grant (2009) points out, bTB has been seen as a 'political disease' ever since a connection was discovered between the spread of bTB from wild badgers to farmed cattle in 1971. Unlike possums, badgers are a native species and a culturally iconic form of wildlife (Cassidy, 2012). During the 1970s and 1980s, attempts to cull badgers as a method of controlling bTB attracted sufficient public concern to deter governments from culling. Instead, government ministers frequently resorted to scientific inquiries as a way of distancing themselves from decision making, the most recent example of which was a ten-year scientific field trial to estimate the effect of badger culling on levels of bTB in cattle (see ISG (2007)).

Public concern over the fate of wildlife remains a key factor in current policy making. It is in this emotionally charged political arena that the Department for Environment Food and Rural Affairs (Defra) has attempted to reform the ways in which bTB is governed. Traditionally, the state has paid for all aspects of bTB control – from conducting disease surveillance, through to compensating farmers. While this approach may have worked when the disease was at a relatively low level, as disease levels have risen so it has become unsustainable. The 2001 FMD crisis had already shown

Defra how expensive managing animal disease outbreaks could be. There was also evidence that the current approach was dysfunctional. For example, generous levels of compensation failed to encourage farmers to implement biosecurity measures to prevent disease. Indeed, in some cases, farmers could be better off suffering from disease than not having had it at all (Campbell and Lee, 2003; Defra, 2003). These financial problems were exacerbated by massive cuts to Defra's own budget in 2010, and a recruitment freeze as part of the UK government's response to the financial crisis. In addition, successive disease outbreaks alerted Defra to the fact that their management was as much a social activity as a veterinary or scientific task. A paternalistic 'we know best' attitude no longer worked: other ways of governing animal disease would have to be found (Enticott *et al.*, 2011).

Defra's response has been to hollow out the state by devolving the costs and responsibilities of disease control to the agricultural industry. In large part, the approach mirrors the New Zealand experience, freeing the state from a wicked problem, while creating a new kind of biosecurity citizenship among the farming industry. Close ties between Great Britain and New Zealand have developed: politicians have visited New Zealand to discover how the Animal Health Board functions, while New Zealand representatives have been invited to the UK to speak to farmers and veterinary experts not just about the disease, but how the AHB is managed by farmers. In justifying the culling of badgers, government ministers have frequently referred to the New Zealand experience, arguing that no country has successfully combatted bTB without addressing bTB in wildlife.

The extent to which this kind of international policy transfer can act as some kind of 'fast policy' solution (Peck, 2002) is questionable. During the 1990s and 2000s, partnership became a key mode of governance for all policy areas in the UK. Animal disease was no exception: Defra's 2005 bTB strategy document referred to 'partnership' twenty-four times, stating that: 'Government intends to work in partnership with stakeholders. This will help us arrive at policy decisions that reflect a robust, fair and cost-effective approach taking into account and balancing stakeholder perspectives' (Department for Environment Food and Rural Affairs, 2005). In 1999, the first bTB Forum was established to discuss options for bTB control other than badger culling. In 2006, Defra established the bTB Advisory Group (TBAG) consisting of a group of core experts from the agriculture and veterinary professions. By 2008 this was replaced by a new group – the Bovine TB Eradication Group for England – which was itself replaced by a new partnership group in 2012 called the Bovine TB Eradication Advisory Group for England. Meanwhile in Wales, a bTB Action Group comprised of agricultural, wildlife and veterinary experts provided advice to government ministers. When this was abandoned in 2008, three new regional eradication groups (REGs) were established, loosely based on New Zealand's system of RAHACs.

Like many partnership arrangements, these attempts failed to provide industry representatives with significant power to shape policy. While some arrangements such as the REGs provided an opportunity to involve farmers in designing attempts to encourage biosecure farming, partnership tended to provide the basis for discussion and communication of government activities to the agricultural industry, rather than a basis for active involvement in policy making. However, the establishment of the Animal Health and Welfare Board for England (AHWBE) in 2011, along with changes to the governance of wildlife controls (see below), suggests a new phase of partnership working. The AHWBE was established following a review into the sharing of costs and responsibilities for disease management between farmers and government (Radcliffe, 2010). The AHWBE's remit is to set policy priorities and recommend the best methods for sharing costs, such as bTB testing, among farmers. Comprised of a range of farming industry representatives and supported by technical staff in Defra, a key role for the AHWBE has been to advise Defra ministers directly on future bTB controls, such as cattle movement restrictions. That the AHWBE directly advises government ministers, rather than officials, represents a new departure, even if the final decision remains with ministers. It remains to be seen how ministers will respond to these changes: despite these new beginnings, there are signs that the politicization of animal disease is ever-present.

A good example of how these new arrangements reflect old styles of governing can be found in recent developments concerning badger culling. In 2008, the Labour Government announced that scientific uncertainty over the effectiveness of badger culling meant that badgers would be instead vaccinated in six areas of England. The strategy relied on creating a market among pest control companies to compete for vaccination contracts that the government would pay for. Contractors, however, were unwilling to bid for this work at reasonable rates, and following the change of administration after the 2010 election, the vaccination scheme was scaled back to just one area. Individuals and private landowners were nevertheless able to apply for badger vaccination licences. For example, the Wildlife Trust began vaccination on its wildlife reserves in Gloucestershire and Shropshire, while the National Trust vaccinated badgers on an estate in Devon. Gradually, what has emerged is a patchwork of individual farmers and large landowners committed to badger vaccination but in a random, uncoordinated fashion.

At the same time, however, the Conservative-led coalition government has also allowed landowners to apply for licences to shoot badgers in an effort to control bTB. To apply for a licence, farmers were asked to form a company that would fund the costs of shooting badgers. In 2012, two areas were selected as pilot areas and culling began in 2013. This approach is interesting in a number of respects. First, the licensing procedure can be seen as the first real step in handing over the costs of disease control to farmers for a disease previously completely controlled by the state. It represents a step towards the neoliberal model of New Zealand and attempts to create biosecure citizenship. Government ministers were clear about this, arguing that disease control was now a matter of choice for farmers: 'if they think it's too expensive, then they won't do it, and it won't happen' (Jim Paice on BBC (2011)). The licensing scheme too was quickly framed in the government's attempts to promote voluntary action – the so-called 'Big Society' approach – across all areas of public service, recalling earlier attempts of relying on the voluntarism of rural citizens (Murdoch, 1997).

Second, the licensing scheme came with its own re-territorialization of disease. As Enticott and Franklin (2009) point out, a common discourse in disease control is the refrain 'disease knows no boundary'. Attempts to manage disease, though, are engaged in constant efforts of de/re-territorialization. In Great Britain, devolution of policy responsibilities has meant that government administrations have had to find ways of re-territorializing disease to legitimize the actions of new governing institutions (Enticott and Franklin, 2009). For the neoliberal approach of New Zealand, disease eradication was tied to national and international spatial scales: the whole purpose of bTB eradication was to ensure the international movement of agricultural products, central to the New Zealand economy and identity. Funding of the Animal Health Board's efforts was drawn proportionately from all farmers throughout New Zealand. Defra's approach, however, territorialized bTB as a local problem to be dealt with by farmers in specific areas. Indeed, this spatialization matched the notion of choice articulated through the licensing procedures. While the licensing procedure was met with general willingness to apply for licences among farmers (a total of twelve applications were received), these arrangements will also mean that its form will vary spatially according to local contexts.

Third, in creating new organizational structures to facilitate badger culling, the boundaries between public and private have become increasingly blurred. The idea that the costs and responsibilities have been handed over wholesale to the private sector is naïve. To begin with, Defra were responsible for establishing the criteria under which any farmers could be awarded a licence. The criteria required upfront payment of the costs of badger culling over four years from all farmers. Based on scientific evidence, Defra also required any licence holders to cull 70 per cent of badger populations in a given area to limit the spread of the disease among surviving populations (known as the perturbation effect, see Woodroffe *et al.* (2006)). Finally, Defra was responsible for significant other costs, including the cost of badger population surveys in the culling zones and the costs of policing protests and defending the policy to a sceptical public.

The organizational challenges of meeting the government's criteria meant that while local farmers fronted the badger cull, there was considerable behind the scenes work by the National Farmers Union (NFU). The NFU saw the success of these pilot schemes as vital for the management of bTB in other areas. In the absence of a national levy or organization such as the AHB to run the pilot culls. The NFU privately filled this breach, contributing resources and expertise to get the job done. But if all that signalled a shift in power from government ministers to the private sector over the choice of whether to kill badgers, there was one final hurdle. In early October 2012, farmers and the NFU decided to delay the two trial cull areas. They did so because new population estimates placed the badger population significantly higher than they thought. The chances of eliminating 80 per cent of badgers in worsening winter weather and mounting public opposition appeared too risky. Instead, farmers decided to wait until the following summer. Defra ministers, however, saw it differently. Not wanting to have appeared to back down to the public or perform another U-turn, the farmers came under intense pressure. In scenes perhaps more reminiscent of past corporative relationships between the Ministry of Agriculture, Fisheries and Food and the NFU, ministers encouraged and demanded that farmers continued with the cull.

In refusing these demands, the delays were interpreted by the media as another government u-turn and a win for a celebrity-led campaign against the cull. But a more accurate interpretation was that there was no political u-turn. Instead, this was the moment when, for the first time in the UK, farmers had taken on responsibility for a statutory animal disease. They had exercised their free choice not to continue the cull for their own sakes. Moreover, the decision was not borne from an idealized neoliberal policy landscape in which the state had ceded responsibility to the private sector, but from a highly complicated organizational regime involving public and private organizations, with some highly visible, while others operated in the half-light. Traditional governance structures and relationships were neither swept away, yet nor were they left fully in the same place. Despite the same purpose of eradicating bTB, this new institutional landscape was quite unlike others in New Zealand. Rather than a coordinated approach to disease control, the UK landscape constructed disease control as an object of competition between those seeking to vaccinate and others wanting to cull.

Conclusion: infected spaces?

The experiences of New Zealand and Great Britain in their attempts to fight bovine Tuberculosis tell us much about the agricultural bioeconomy, its spatial governance and the kinds of natures and societies they enact. In conclusion, it is worth reiterating some key points.

First, just like disease, the bioeconomy's neoliberal response to biosecurity threats have become infectious, spreading between countries and continents. While this may reflect how the range of acceptable policy options has narrowed under neoliberal regimes (Peck, 2010), for biosecurity we can begin to trace an international movement of neoliberal responses. New Zealand's response to bTB mirrored elements of Australia's successful eradication campaign (see Lehane, 1996), partly owing to physical proximity but also because it chimed with the economic crisis facing New Zealand in the mid 1980s. Recently, governments in Great Britain have begun to draw strongly on these ideas, not just because of its own economic crisis, but the international mobility of vets has also helped establish these neoliberal techniques as part of the techniques of biosecurity, as much as biosecurity practices and technologies (see, for example, More, 2007). Importantly, though, these ideas do not always travel well, and as we have seen, require translation by local actors if they are to work.

Second, the chapter has shown how biosecurity is constructed at different spatial scales, yet interactions between each scale are vital to biosecurity working. In New Zealand, biosecurity techniques practised at the local scale – on farms and woodlands – were made possible by articulating a national sense of disease responsibility and linking biosecurity with symbols of national identity, which themselves were linked to an international space of economic and agricultural flows that was protected by subscribing to international biosecurity rules and practices. The complementarity between these spatial scales allowed biosecurity practices to work, in that there were no conflicts that could not be overcome. At the same time, they enacted particular versions of nature and disease. In New Zealand, for example, the involvement of farmers in decision making has changed the way disease is understood. In England by contrast, the lack of complementarity between the different scales of biosecurity has meant that proposed solutions have constantly stalled. The role of badgers in national culture and identity has meant that badger culling has met with stiff public resistance, even among farmers, when governments have attempted it. Attempts to promote a national biosecurity citizenship have similarly come unstuck – either because the failure to instigate badger culling has led to disinterest and distrust among farmers, or because politicians feel unable to let farmers construct their own biosecurity futures. Unlike New Zealand there is no perceived threat to agricultural exports, nor do they occupy a significant proportion of GDP – reflecting the vague and complex character of biosecurity relations found in all spatial scales. In turn, the lack of fit between spatial scales points to the difficulties of biosecurity policy transfer. While governments may like to laud the efforts of others to justify controversial biosecurity policies – such as wildlife control – reaching for biosecurity solutions developed elsewhere is likely to fail unless they can be significantly translated to local conditions.

Finally, if biosecurity is a way of enacting the neoliberal bioeconomy, then its form is not the same everywhere. Neoliberalism's reach into the governance of biosecurity has been anything but universal with local contexts and actors shaping its appearance. In the context of New Zealand's rolling back of the state and withdrawal of agricultural subsidies, the establishment of the farmer-levied Animal Health Board might seem unsurprising. But while the

AHB emerged at a time of economic crisis, it did not emerge overnight. The AHB is rooted in a much deeper historical context in which the Crown played a significant role: the AHB can be seen at the end of a trajectory that had been developing for forty years. In the UK, by contrast, a more chaotic structure to biosecurity governance has emerged rapidly in which different methods of disease control are in competition in an ever-changing institutional landscape of biosecurity. It is as if neoliberal solutions have gone looking for biosecurity problems in a desperate attempt to appear to be doing something about a problem, whether or not there is conclusive evidence that they work (cf. Cohen *et al.*, 1972; Prince, 2012).

In doing so, biosecurity helps the bioeconomy draw a veil across certain forms of agriculture and alternative biosecurity regimes. Yet while the bioeconomy may fear the effects of mobile and uncontrollable natures, it is also responsible for creating them. Other versions of agriculture might pay greater attention to alternative biosecurity regimes and question the extent to which old diseases (such as bTB) have been surpassed by new ones; whether biosecurity regulations are the cause of problems rather than their solution; and whether local versions of biosecurity are more realistic and attractive than adherence to international rules. At the same time, it will be upon already marginal farms – many of which do not share the bioeconomy's reliance on agricultural mobility – that the costs of this version of agriculture will fall. It may be that the bioeconomy's drawing of biosecurity will be eroded by wider social change in the countryside. As new populations move in with different attitudes towards nature, long-standing biosecurity solutions such as wildlife control may become increasingly difficult, leading to a broader debate as to what constitutes biosecure citizenship. Perhaps, as Law (2006) suggests, the real failure of biosecurity will be to ignore the systemic reasons for the spread of animal disease and fail to consider alternative ways of conceptualizing biosecurity.

References

Animal Health Board (2012) *Annual Report 2011/12*, Wellington: Animal Health Board.

Barker, K. (2008) 'Flexible boundaries in biosecurity: Accommodating gorse in Aotearoa New Zealand', *Environment and Planning A*, vol. 40, no. 7, pp. 1598–1614.

Barker, K. (2010) 'Biosecure citizenship: Politicising symbiotic associations and the construction of biological threat', *Transactions of the Institute of British Geographers*, vol. 35, no. 3, pp. 350–363.

BBC (2011) *Countryfile*, television programme, 11 September, London: BBC.

Bennett, K. and Phillipson, J. (2004) 'A plague upon their houses: Revelations of the foot and mouth disease epidemic for business households', *Sociologia Ruralis*, vol. 44, no. 3, pp. 261–284.

Braun, B. (2007) 'Biopolitics and the molecularization of life', *Cultural Geographies*, vol. 14, no. 1, pp. 6–28.

Brenner, N., Peck, J. and Theodore, N.I.K. (2010) 'Variegated neoliberalization: Geographies, modalities, pathways', *Global Networks*, vol. 10, no. 2, pp. 182–222.

Busch, L. (2010) 'Can fairy tales come true? The surprising story of neoliberalism and world agriculture', *Sociologia Ruralis*, vol. 50, no. 4, pp. 331–351.

Campbell, I.D. and Lee, R. (2003) 'Carnage by computer: The blackboard economies of the 2001 foot and mouth epidemic', *Social and Legal Studies*, vol. 12, no. 4, pp. 425–459.

Cassidy, A. (2012) 'Vermin, victims and disease: UK framings of badgers in and beyond the bovine TB controversy', *Sociologia Ruralis*, vol. 52, no. 2, pp. 192–214.

Cohen, M., March, J. and Olsen, J. (1972) 'A garbage can model of organizational choice', *Administrative Science Quarterly*, vol. 17, no. 1, pp. 1–25.

Convery, I., Mort, M., Baxter, J. and Bailey, C. (2008) *Animal Disease and Human Trauma: Emotional geographies of disaster*, London: Palgrave Macmillan.

Davidson, R.M. (2002) 'Control and eradication of animal diseases in New Zealand', *New Zealand Veterinary Journal*, vol. 50, no. 3 (supplement), pp. 6–12.

Department for Environment Food and Rural Affairs (2003) *Assessment of the Economic Impacts of bTB and Alternative Control Policies: Final project report SE3112*, London: Defra.

Department for Environment Food and Rural Affairs (2005) *Government Strategic Framework for the Sustainable Control of Bovine Tuberculosis (bTB) in Great Britain*, London: Defra.

Dibden, J., Higgins, V. and Cocklin, C. (2011) 'Harmonizing the governance of farming risks: The regulation and contestation of agricultural biosecurity and biotechnology in Australia', *Australian Geographer*, vol. 42, no. 2, pp. 105–122.

Donaldson, A. (2008) 'Biosecurity after event: Risk politics and animal disease', *Environment and Planning A*, vol. 40, no. 7, pp. 1552–1567.

Enticott, G. (2008) 'The ecological paradox: Social and natural consequences of the geographies of animal health promotion', *Transactions of the Institute of British Geographers*, vol. 33, no. 4, pp. 433–446.

Enticott, G. (2011) 'Techniques of neutralising wildlife crime in rural England and Wales', *Journal of Rural Studies*, vol. 27, no. 2, pp. 200–208.

Enticott, G. and Franklin, A. (2009) 'Biosecurity, expertise and the institutional void: The case of bovine Tuberculosis', *Sociologia Ruralis*, vol. 49, no. 4, pp. 375–393.

Enticott, G. and Wilkinson, K. (2013) 'Biosecurity: Whose knowledge counts?', in A. Dobson, K. Barker and S.L. Taylor (eds) *Biosecurity: The socio-politics of invasive species and infectious diseases*, London and New York: Routledge, pp. 91–104.

Enticott, G., Donaldson, A., Lowe, P., Power, M., Proctor, A. and Wilkinson, A. (2011) 'The changing role of veterinary expertise in the food chain', *Philosophical Transactions of the Royal Society B: Biological Sciences*, vol. 366, pp. 1955–1965.

Fish, R., Austin, Z., Christley, R., Haygarth, P.M., Heathwaite, L.A., Latham, S., Medd, W., Mort, M., Oliver, D.M., Pickup, R., Wastling, J.M. and Wynne, B. (2011) 'Uncertainties in the governance of animal disease: An interdisciplinary framework for analysis', *Philosophical Transactions of the Royal Society B: Biological Sciences*, vol. 366, no. 1573, pp. 2023–2034.

Grant, W. (2009) 'Intractable policy failure: The case of bovine TB and badgers', *British Journal of Politics and International Relations*, vol. 11, no. 4, pp. 557–573.

Higgins, V. and Dibden, J. (2011) 'Biosecurity, trade liberalisation, and the (anti)politics of risk analysis: The Australia–New Zealand apples dispute', *Environment and Planning A*, vol. 43, no. 2, pp. 393–409.

Higgins, V. and Larner, W. (2010) 'Standards and standardization as a social scientific problem', in V. Higgins and W. Larner (eds) *Calculating the Social: Standards and the reconfiguration of governing*, Basingstoke: Palgrave Macmillan, pp. 1–17.

Higgins, V., Dibden, J. and Cocklin, C. (2012) 'Market instruments and the neoliberalisation of land management in rural Australia', *Geoforum*, vol. 43, no. 3, pp. 377–386.

Hinchliffe, S., Allen, J., Lavau, S., Bingham, N. and Carter, S. (2013) 'Biosecurity and the topologies of infected life: From borderlines to borderlands', *Transactions of the Institute of British Geographers*, vol. 38, no. 4, pp. 531–543.

Hodge, I.D. and Adams, W.M. (2012) 'Neoliberalisation, rural land trusts and institutional blending', *Geoforum*, vol. 43, no. 3, pp. 472–482.

Ilbery, B. (2012) 'Interrogating food security and infectious animal and plant diseases: A critical introduction', *The Geographical Journal*, vol. 178, no. 4, pp. 308–312.

Independent Scientific Group (ISG) (2007) *Bovine Tuberculosis: The scientific evidence*, London: Defra.

Jamieson, S. (1960) 'Bovine tuberculosis: Problems and prospects of eradication', *N.Z. Department of Agriculture*, Bulletin, no. 361, pp. 1–8.

Law, J. (2006) 'Disaster in agriculture: Or foot and mouth mobilities', *Environment and Planning A*, vol. 38, no. 2, pp. 227–239.

Lehane, R. (1996) *Beating the Odds in a Big Country. The eradication of bovine brucellosis and tuberculosis in Australia*, Collingwood, Australia: CSIRO.

Lockwood, J.L., Cassey, P. and Blackburn, T. (2005) 'The role of propagule pressure in explaining species invasions', *Trends in Ecology and Evolution*, vol. 20, pp. 223–228.

MAFF (1965) *Animal Health – a Centenary 1865–1965*, London: HMSO.

Maye, D., Dibden, J., Higgins, V. and Potter, C. (2012) 'Governing biosecurity in a neoliberal world: Comparative perspectives from Australia and the United Kingdom', *Environment and Planning A*, vol. 44, no. 1, pp. 150–168.

More, S. (2007) 'Shaping our future: Animal health in a global trading environment', *Irish Veterinary Journal*, vol. 60, no. 9, pp. 540–545.

Mort, M., Convery, I., Baxter, J. and Bailey, C. (2005) 'Psychosocial effects of the 2001 UK foot and mouth disease epidemic in a rural population: Qualitative diary based study', *British Medical Journal*, vol. 331, no. 7527, pp. 1234–1237.

Murdoch, J. (1997) 'The shifting territory of government: Some insights from the rural white paper', *Area*, vol. 29, no. 2, pp. 109–118.

Nerlich, B. and Wright, N. (2006) 'Biosecurity and insecurity: The interaction between policy and ritual during the foot and mouth crisis', *Environmental Values*, vol. 15, no. 4, pp. 441–462.

Parliamentary Commissioner for the Environment (2011) *Evaluating the Use of 1080: Predators, poisons and silent forests*, Wellington: Parliamentary Commissioner for the Environment.

Pearce, F. (1996) 'The aliens have landed', *The Scotsman*, 1 July 1996.

Peck, J. (2002) 'Political economies of scale: Fast policy, interscalar relations, and neoliberal workfare', *Economic Geography*, vol. 78, no. 3, pp. 331–360.

Peck, J. (2010) *Constructions of Neoliberal Reason*, Oxford: Oxford University Press.

Prince, R. (2010) 'Policy transfer as policy assemblage: Making policy for the creative industries in New Zealand', *Environment and Planning A*, vol. 42, no. 1, pp. 169–186.

Prince, R. (2012) 'Policy transfer, consultants and the geographies of governance', *Progress in Human Geography*, vol. 36, no. 2, pp. 188–203.

Radcliffe, R. (2010) *Responsibility and Cost Sharing for Animal Health and Welfare: Final report*, London: Defra.

Waage, J.K. and Mumford, J.D. (2008) 'Agricultural biosecurity', *Philosophical Transactions of the Royal Society B: Biological Sciences*, vol. 363, no. 1492, pp. 863–876.

Waddington, K. (2004) 'To stamp out "so terrible a malady": Bovine tuberculosis and tuberculin testing in Britain, 1890–1939', *Medical History*, vol. 48, no. 1, pp. 29–48.

Ward, N., Donaldson, A. and Lowe, P. (2004) 'Policy framing and learning the lessons from the UK's foot and mouth disease crisis', *Environment and Planning C: Government and policy*, vol. 22, no. 2, pp. 291–306.

Wilkinson, K. (2011) 'Organised chaos: An interpretive approach to evidence-based policy making in Defra', *Political Studies*, vol. 59, no. 4, pp. 959–977.

Williamson, M. (2006) 'Explaining and predicting the success of invading species at different stages of invasion', *Biological Invasions*, vol. 8, no. 7, pp. 1561–1568.

Woodroffe, R., Donnelly, C.A., Cox, D.R., Bourne, F.J., Cheeseman, C.L., Delahay, R.J., Gettinby, G., McInerney, J.P. and Morrison, W.I. (2006) 'Effects of culling on badger Meles meles spatial organization: Implications for the control of bovine tuberculosis', *Journal of Applied Ecology*, vol. 43, no. 1, pp. 1–10.

7 Improving animal welfare in Europe

Cases of comparative bio-sustainabilities

Mara Miele and John Lever

Introduction: animal welfare as an element of agricultural sustainability?

In a recent book chapter, Buller and Morris (2009) have interrogated the problematic nature of the relationship between farm animals and agricultural sustainability.[1] Specifically, they ask whether the agricultural sustainability agenda serves the 'interests' of farm animals? Does it provoke a different, more ethical, relationship between farm animals and society than is the case in debates surrounding 'conventional' agriculture? Or does it merely serve to reinforce existing, and largely instrumental, animal–society relations, albeit locating them within a different moral framework? These authors trace the debate about farm animals and sustainability and identify a first discourse that conceives farm animals as a threat to sustainability, and a second that sees animals as vectors for delivering sustainability. In recent years, they underline how these two conceptualizations have been joined by a new element in the incorporation of farm animals within conceptions of sustainability. This sees farm animals as the targets of sustainability, as notions of animal welfare become integrated into the very definitions of sustainable agriculture.

They conclude by arguing that the shift from representing animals as threats to sustainability to the current position where animal welfare is included within discourses of agricultural sustainability has not fundamentally reconfigured the relationship between human and nonhuman animals; 'sustainable agriculture is not vegan, but it can be caring' (Buller and Morris, 2009, p. 141). However, they acknowledge that the welfarist position has brought about some significant effects, by making nonhuman animals visible in their complexity, and exposing or deconstructing the relations that humans create for them, thus helping to facilitate the emergence of less instrumental human/nonhuman relations, especially in farming practices. We would argue that it also promoted a different moral standing of animals, as clearly indicated in the European Union Amsterdam Treaty (1997), where nonhuman animals are now recognized as *sentient beings* and not simply as a means to an end (e.g. inputs for production of food, in the case of farm animals).

Starting from the acknowledgement of the effects of '*an animal welfarist*' position in promoting changes in human–animal relations and ideas about sustainable farming, in this chapter we want to address the complexity of the initiatives for improving animal welfare in Europe, the difficulties in implementing common policies and the current experimentations of new instruments for overcoming the limits of the traditional regulatory approach adopted until now (see, for example, the European Animal Welfare Strategy 2012–2016) by the European Union.

The chapter proceeds in the following way: first, we will point to the growing public concern for animal welfare in Europe (see Miele and Evans, 2010; Evans and Miele, 2012), now perceived as an integral part of the civilization process in Europe and an essential element of a new model of sustainable agriculture in the EU regulation (Miele and Lever, 2013). Here we see that the protection of animals has long been considered a common value of European citizens. The European Union has identified the well-being of farm animals as a priority in EU legislation and has achieved world leadership with the many initiatives in the last forty years (Blokhuis *et al.*, 2013). However, even with a significant legislative effort for improving farm animal welfare, the public concerns for the quality of life of farm animals has grown in the last three decades and this is especially linked to the intensification and industrialization of animal farming and to the BSE and FMD epidemics (European Commission, 2007). Therefore, we argue that these sensibilities and public concerns have promoted a variety of interventions both at regulatory level and a series of more diffuse, bottom-up, market initiatives that reflect the different interpretations of animal welfare (e.g. which aspects of animals' lives are more important) and different course of actions for improvements in it. Drawing on the work by Kjærnes, Bock and Miele (2009) we, secondly, refer to various models of institutional settings that address the differences in Europe in terms of socio-economic conditions for the development of the animal farming industry, the market for animal friendly labelled products, the style of government as well as regional, historical and cultural factors. We then conclude with some reflections on the marketization of animal welfare and the possible advantages and limitations of the current focus on the emerging '*(super)market* mechanism' for improving animals' welfare, as indicated in the European Animal Welfare Strategy 2012–2016.

Animal welfare: a multidimensional concept

In Europe we can identify two main mechanisms of animal welfare improvement: initiatives in terms of *regulation of animal farming by law*, both at supranational and national levels and initiatives as part of *market differentiation* for animal products (a form of bottom-up market governance) (see Miele *et al.*, 2005, 2013). The first mechanism's aim, regulating animal farming by law, is based on the definition of a 'minimum standard' of animal welfare looking at the 'resources' available to farm animals (e.g. minimum space for crates or cages). But the existing legislation does not cover all types of production or all the farmed animal species. Moreover, the extent to which the supranational regulation is implemented and monitored varies across EU countries and national regulations also vary. This has created a space for the second mechanism of improvement that has been developed more recently in association with various types of quality assurance and market differentiation schemes (Miele *et al*., 2005; Buller and Roe, 2011, 2013). Recent studies have pointed out that in European countries there are major differences in the social division of responsibility for farm animal welfare, and there are thus significant differences in who is promoting such initiatives, that is, whether they emerge from producer groups and processors, big retailers, alternative suppliers, NGOs, or public authorities.

The emergence of market strategies promoted by various actors in animal supply chains is strongly interdependent and they are affected by the institutional settings in which animal welfare issues are addressed. Kjærnes *et al*. (2009) identify three models in Europe: the *(super)market model* (more evident in the UK and the Netherlands), the *welfare state model* (more evident in Norway and Sweden) and the *terroir model* (developed in France and Italy). We propose to add a fourth one, the '*no issue*' model, that exemplifies the minimal condition of several EU countries where there is little debate about animal welfare and there are very limited initiatives aside compliance with the EU regulations. These models are decisive for communication about the welfare status of the animal products to the end consumers, as well as for the ways in which consumers are expected and are empowered to choose on the market, that is, they are important for understanding the different level of political consumerism about animal welfare in different countries and the conditions for the emergence of the 'animal friendly consumer'. As Stolle *et al*. (2005) have argued, political consumers opt for specific producers or goods because they want to intervene in institutional or market practices. Their purchases are based on considerations about values such as justice or fairness, or on an assessment of industry and government practices. 'Regardless of whether political consumers act individually or collectively, their market choices reflect an understanding of material products as embedded in a complex social and normative context, which can be called the politics behind products' (Stolle *et al*., 2005, p. 246). But the conditions for political consumerism to emerge imply that the politics behind the products needs to be brought into public debate (Marres, 2009) and into the marketplace, and, crucially, that a significant number of consumers have the means to act upon their ethical or political preferences.

We will explore these different institutional arrangements with reference to comparative empirical studies in particular member states in the light of the recent debate on political consumerism (Stolle *et al*., 2005; Koos, 2012) in order to explore the presence or absence of the conditions that we mentioned above for the emergence of both the market for animal friendly products and the 'animal friendly consumer'.

Animal welfare institutional settings models

The 'no issue' model: Hungary

Across Europe there are different political and public framings of farm animal welfare issues. In Hungary animal welfare is not a commonly used term. According to Fodor and Redai:

> [W]hile there has been an emergence of alternative social and cultural forms since the 1990s, the most notable change is the increase in social inequalities to a point not seen in Hungary in generations. Class inequalities have become prominent over the past 15 years, with poverty on the rise, the middle class losing ground, and a relatively small elite enjoying economic success and security. Social and economic insecurity came as a novel experience for many people after decades of state socialism. While nowadays class inequalities in Hungary are smaller than in most other EU countries according to the GINI coefficient, the change from 1989 is vast and people are still adjusting. Inequalities grew especially fast until the mid 1990s when they slowed and, some argue, have decreased slightly since. In addition, as mentioned above, the government gives preference to what they call 'economic development' over social justice, and non-governmental organizations are too weak to fight effectively for the downtrodden.
>
> (2006, p. 38)

Concerns for animal welfare among Hungarian actors and their interpretations assigned to the definition of animal welfare are thus highly variable. Whereas government actors seem to define animal welfare in terms of compliance with EU animal welfare directives, the Fauna Society, a non-profit organization, affiliated to the World Society for the Protection of Animals (WSPA), prioritizes animal cruelty and focuses on non-farming animals. It is often argued that prevailing worries over economic uncertainty and low purchasing power do not leave much space for other types of concerns, such as animal welfare, while shopping for food. Hungarian consumers expect farmers, transporters and slaughterhouses to take on responsibility for the ethical treatment of farm animals. They do not assume personal responsibility for animal welfare and are pessimistic that they can influence change and expect state protection through the use of regulation. They are disillusioned by poor enforcement of the law, believing that greater enforcement of current animal welfare regulations rather than further regulation is needed (McIntyre and Cowan, 2007).

The terroir model: France and Italy

In several other European countries animal welfare issues have not been given priority in the public agenda. In Italy for example, environmental protection, social equity and animal welfare emerged as new social problems during the 1990s, and more so from the beginning of 2000, but they never achieved the political relevance or media attention as in the Netherlands or the UK. Owing to the prominence given by the Italian media to food scandals in the 1980s and 1990s (Miele and Parisi, 2000), the general public concern regarding food supply has increased and it is within this context that animal welfare issues were brought for the first time into the public debate. The debate resulted in more attention about food safety and healthiness rather than more awareness about farming practices or animal welfare issues. Moreover, in Italy, there are large NGOs (LAV, anti-vivisection league) whose philosophy is centred on an 'animal rights' or abolitionist perspective (Francione and Gardner, 2010). They mostly focus on the treatment of companion animals (e.g. dog abandonment) and they advocate vegetarianism and veganism by largely disregarding animal welfare improvement. Their initiatives are strongly associated with the internet even though they are also active in lobbying policy makers at national and local level (Miele *et al.*, 2007).

With the exception of those committed actors within the animal welfare movement, animal welfare does not emerge as a major concern among French citizens. There is, nevertheless, a gradual shift in French attitudes and a growing awareness of welfare issues, prompted partly by the emergence of animal welfare organizations, by the recent food scares, but also by a dramatic growth of quality labelling, including organic, and the inclusion of welfare conditions within quality production criteria (Buller and Cesar, 2007; Ingenbleek and Immink, 2011). The food crises of the 1990s and early twenty-first century have also had a major impact upon marketing and commercial strategies. A great deal of information is now available to consumers on the feed given to farm animals, on the length of life (particularly for poultry, far less so for beef cattle (Poulain *et al.*, 2007). Animal welfare issues are strongly bundled with concepts of tradition and typicality in the communication of quality characteristics to consumers (e.g. Label Rouge) and very seldom are addressed as single issues (Buller and Cesar, 2007).

The welfare state model: Scandinavia

According to Kjærnes and Lavik (2005), Swedes value animal welfare the highest (together with the Hungarians), but are not as worried as consumers in many other countries. Generally, a majority think that animals are treated quite well in Sweden and as long as the food is of Swedish origin people tend to believe that animal welfare is taken care of. In a way, Sweden is perceived as the animal paradise; Swedes value animal welfare but don't see the point of doing more than is already done – in Sweden. Still, attention appears to have increased in recent years and the question of animal welfare and, in particular, animal rights has become an everyday and justifiable issue. Even on the formal political agenda, in the Parliament, government and political parties, animal welfare concerns have become an issue. A number of non-governmental organizations in Sweden focus on animal welfare, animal rights, and/or the protection of animals, some of them with a considerable number of members (Petterson and Bergman, 2007).

Even less concern is found in Norway. There is widespread consensus that Norwegian farm animals have a far better life than animals bred in other countries. In particular, the protection of small-scale production and farmers' standard of living are regarded as important strategies for assuring good animal welfare. Good farmer welfare (supported by the state) is seen as important for promoting animal welfare and animal welfare problems are often presented as social tragedies among farmers. As in Sweden, there is a widespread perception that animal welfare issues in Norway are well addressed by public regulations, and there are no civic society calls for different instruments or initiatives to be explored. Organizations bringing up issues of animal welfare and animal rights give relatively little attention towards the specific issues of the welfare of farm animals. Moreover, as publicly initiated efforts are seen to be most efficient, NGOs have, on issues related to farm animals, directed their efforts mainly towards public decision-making processes, rather than mass mobilization and shopping strategies (Terragni and Turjusen, 2007).

The supermarket model: the UK and the Netherlands

In some countries, most prominently the Netherlands and the UK, animal welfare is an established and more explicit issue on the political agenda, and it is in these countries that we see more clearly the emergence of a '*(super) market model*'. The Dutch political debate about farm animal welfare has been promoted mainly by NGOs and the recent entry of the Party for Animals (2006) in parliament, has further pushed animal welfare issues in the political debate. This civic engagement has resulted in changes in laws and regulations but has also influenced market practices. One example is the ban on battery eggs: since 2004 Dutch supermarkets sell eggs produced in alternatives to caged systems. Some retailers and multinationals, in addition, have decided to stock and sell only free-range eggs in their processed foods (Schipper *et al.*, 2007).

In the UK, public regulation of animal farming and market initiatives seem to coexist consensually. While there is much public debate around the welfare of farm animals, and the regulation of animal welfare is, in certain cases, higher than the European one, increasing attention has been directed towards market initiatives. The priority area of investment has been in the development of national assurance schemes, a term that covers and includes farm assurance, that is, on-farm practices, and quality assurance, namely, supply chain and farm assurance guaranteed practices. As Susanne Freidberg (2004) has pointed out, here the media and the powerful animal welfare NGOs have been most effective in engaging retailers in a dialogue and a search for better farm animal welfare standards, promoting changes in farming practices. These are often resented by the more conservative part of the farming community that predominantly see these initiatives as driven by dubious 'urban consumers' wishes and beliefs, while the more dynamic and market-oriented farmers see it as an opportunity for market differentiation (Mayfield *et al.*, 2007; Buller and Roe, 2013; Miele and Lever, 2013).

The emergence of the *(super)market* model for improving animal welfare

While some of the basic legislation on animal welfare is harmonized in Europe, the role of public authorities in animal welfare regulation is highly variable and changing. At least three different positions can be identified; a *politicized*, but also market-oriented approach, a *universal* or protective approach, and a more *fragmented*, uncoordinated approach (Kjærnes, 2012). There are a variety of schemes and labels addressing animal welfare in the UK, but the background and context are very different, and the situation far less complex compared to France. British assurance schemes have been integral to retailer-led structural changes to the meat/dairy/egg supply chains, in which many of the competencies between slaughter and packing have been incorporated into single firms. Each retailer usually has only two to three 'key' suppliers (sometimes only one) that source meat according to retailers' specifications. The three major retailers with a combined market share of over 60 per cent source all their fresh meat, dairy and egg products from quality assured suppliers and producers (Murdoch, 2005; Roe and Murdoch, 2006).

Given this high concentration in the food retailing system, producers primarily see assurance schemes as guaranteeing market access. Large retailers use the schemes in two interlinked ways – to protect their brand by ensuring integrity of their products and communicating this to consumers, and differentiating product ranges. The four main assurance schemes in the UK are Assured Food Standards (AFS) (an umbrella industry standard for the species-specific schemes), Freedom Food (welfare-specific assurance scheme from RSPCA), the Soil Association organic standard, and the organic farmers and growers standard, who have developed different roles in the market. AFS is now an industry standard: it guarantees market access rather than secures a market premium. Retailers generally use the AFS logo (a red tractor against a British flag) on all products that meet the standards. Organic standards and logos are used to differentiate organic ranges and have been very successful in securing a market premium. This is reflected by producers' motivations for working with organic standards, which include added premiums. Freedom Food has occupied a position between the organic and the industry standard, while large retailers, pioneered by Waitrose and Marks & Spencer, have started looking beyond 'generic' assurance schemes to more 'bespoke' production systems (especially for 'added value' tiers) that reflect their brand, and its ethical integrity (Roe and Murdoch, 2006).

In order to add value to animal products, British retailers do, in some circumstances, use welfare claims on packaging. This information is not there to 'educate' but to attract the consumer, and it endorses folk notions of farm animal welfare, evoking 'naturalistic' pictures of farmed animals associated with wider notions of quality: tradition, taste, environmental concerns, etc. (see Roe and Murdoch, 2006). Moreover, large retailers seem keen to use their brand as shorthand for a whole host of ethical considerations, backed up by detailed information on websites, audits by NGOs, and Corporate Social Responsibility reports. This of course allows them a certain flexibility in terms of defining farm animal welfare. They are wary of generic standards and prefer to have the ability to differentiate products, and thus their particular brand, using animal welfare as *one* of the criteria in a range of quality attributes; they also recognize the difficulty in conveying to consumers the detail of welfare improvements (Buller and Roe, 2011).

Until the entrance of the Party for Animals, the Dutch government and farmers' organizations were both in favour of organizing farm animal welfare improvements through the market. Among retailers this was also a favoured strategy, since it allowed them to create niche markets for products with higher animal welfare standards and the possibility to gain more profits. The key point here is that corporate retailers see animal welfare conventions as a means to add value. Market actions still play a significant role and several Dutch initiatives are based on cooperation between farmers, NGOs, the ministry of agriculture and retailers. Some of them are quite established, such as the free-range production schemes for pig, beef and chicken meat that were initiated around 1985 by animal protection NGOs, the consumer association, and the Ministry of Agriculture and initially hosted by the PVE (the product boards of livestock, meat and eggs). Another scheme is 'Agro-milieukeur' which initially focused on environmentally friendly production. Recently also, boar meat from un-castrated pigs is produced and sold under this label. Other recent initiatives taken by the animal protection board in cooperation with retailers and the farmers' union concern the design of animal friendly housing systems (comfort stable) for pigs and, recently, laying hens and the development of a new, slow-growing chicken breed that was introduced in several supermarkets under the label of Volwaerd chicken. There are now similar initiatives for pigs (Schipper *et al.*, 2007).

In contrast to the high level of private initiatives in these two countries (that show a diversity of schemes and labels), market and competitive issues are less evident for animal welfare in other countries. The large retailers in Italy are highly sensitive to consumer opinions and preferences and they focus their initiatives on food quality and safety rather than strictly ethical values such as animal welfare. The retailers claim that consumers are not (or not yet) expressing their concerns for animal welfare in their food choices; in other words, 'perceived quality' sells better than the ethical concern for the welfare of animals. Still, in Italy, traditional butchers sell almost 50 per cent of all meats in outlets where there is a strong emphasis on the organoleptic quality (e.g. taste) of animal products rather than branding or explicit labels. Some retailers do offer a range of animal welfare friendly products (NaturaSì, Esselunga, CoopItalia, Conad), some offer a few (Despar, Proda, Sigma, Standa, GS-Carrefour) while others do not sell any animal welfare friendly labelled products (e.g. Lidl); none of the retailers has a label dedicated only to animal welfare. Supermarket own-labels tend to concentrate on other issues, such as food safety, traceability, controlled supply chains, organic production, or typical regional products. At this stage, the limited number of welfare friendly products on the market are supplied and branded by the food industry. In any case, animal welfare is not yet used as a strategic element, but it is starting to appear alongside hygiene and health standards, taste and other organoleptic characteristics of the product. The conditions for farmers to join specific animal welfare production schemes are linked either to the need for maintaining market access (e.g. the need to comply with supermarket requirements, as in the case of CoopItalia or NaturaSì), or by changing the type of production as in the case for niche products (e.g. organic), for which consumers are ready to pay a premium price (Miele and Parisi, 2000; Miele *et al.*, 2007).

'*Safe and sound, not harming animals*' is the way Swedish suppliers generically describe animal welfare, and safe and sound harmonize with governmental regulations. Because of this, suppliers don't find any competitive advantage in introducing private labels referring to animal welfare. This 'market' has not been created. The animal welfare segment is too small and it is not even perceived as an independent concept that can be used in order to build and gain competitive advantages. Even if the consumer does want to know, they do not want to pay. Many industry representatives believe that consumers do not want detailed information about animal production when buying food. Marketing strategies that dissociate meat from live animals represent a way of meeting such tendencies. Programmes including animal welfare criteria, though not specifically on animal welfare, have been created and run by Swedish producers' organizations (e.g. Swedish meats, Swedish milk, Swedish poultry). A variety of quality assurance schemes for cattle, pigs and poultry cover animal welfare components, addressing animal health, hygiene, technical standards for buildings and facilities, but also aspects of cropping, forestry and environment. There are specific organic schemes, but many organic producers also participate in schemes in which even conventional producers participate. The farmers have a limited choice between schemes to specify the standard of production quality, but there is no realistic option *not* to join a scheme – so few farmers do this. Assurance companies, slaughterhouses, dairies, egg suppliers, and processing companies often demand participation in certain schemes as a precondition for buying their animals or products. To some degree, Swedish retailers have started using assurance schemes.

Co-op is using KrAv (the organic certifier) to certify eggs and ICA is using free-range eggs for their own-label products. But with the exception of organic schemes and Svensk Fågel (a broiler production scheme) there are no schemes for producers that involve product labels for consumers. Retailers work with consumer labels of their own, also regional ones, but the correspondence between producer schemes and consumer labels is not obvious (Petterson and Bergman, 2007).

Norway places, relatively speaking, little emphasis on assurance schemes and, of those that do exist, many represent public–private initiatives. The most widespread scheme is 'the quality system for agriculture' (Ksl), introduced in 1993 upon the initiative of agricultural authorities, together with farmers' organizations and the farmer-owned cooperative manufacturers. Even if it is voluntary, very few farmers do not participate in this scheme (and receive a – publicly regulated – premium). Ksl does not enhance specific better welfare practices, as the scheme in this regard refers to public regulation, but does have a role in guiding the overall implementation of standards via a quality assurance system. Norwegian producers are not enthusiastic about differentiating schemes. Product claims of high animal welfare are seen as equal to communicating that the rest is produced with lower standards. Instead, a uniform and high quality production is regarded as preferable. Organic production is certified by Debio, a privately owned agency working with authority delegated by the Ministry of Agriculture and the national food authority. Unlike Ksl, the Debio standard has resulted in a product label, the Ø-label. To some extent, Debio has stricter regulations with respect to animal welfare, but this was not actively marketed until a campaign launched in 2006 ('*pampered food*'). There are particular industry initiatives, mainly in the form of educational programmes. An example is the plan of Action for Animal Welfare of 2001 by the organization for pig producers, Norsvin. An animal protection group (Dyrebeskyttelsen) acted as an advisor in the shaping of this training programme. Explicit use of animal welfare in marketing is uncommon, but there are some exceptions. The egg supplier Norgården used the slogan '*we take animal welfare seriously*' on the packaging of their eggs. This producer was the first to introduce organic eggs and eggs from free-range systems to the market. Norgården is now owned by the cooperative processing giant GildePrior. Another example is 'Grøstad Gris', who use animal welfare most actively as a communication strategy with consumers (Terragni and Turjusen, 2007).

The overall supply of welfare labelled food products is very limited in Hungary. Moreover, the general demand for meat decreased in the 1990s, with beef and pork consumption per person/per year showing a sharp decline while the consumption of poultry increased to a higher level than the average in the EU-15. Beef consumption is quite low while pig meat consumption has decreased from a high level in earlier years to the level of the EU-15. Milk consumption in Hungary also fell during the 1990s (IAMO, 2004 in McIntyre and Cowan, 2007, p. 4).

A small number of production schemes can be identified that, explicitly or implicitly, address animal welfare issues. In addition to organic production, there are some small-scale local schemes in place for, for example, barn eggs. There are some broader schemes and labels that many associate with good production and high quality. The trademark Quality Food from Hungary is based on guarantees that the ISO[2] standard for environment-oriented management is attained and is based on an integrated quality management system. These guarantees do not include animal welfare today, but might do so in the future. The low significance of animal welfare schemes and labels should be understood within the context of a food market with a low proportion of pre-packaged and labelled products sold in supermarkets. The general focus in food production is on safety and efficiency and adherence to legal standards. The Hungarian multiples have not been generally supportive of welfare friendly products mainly because of associated high prices and poor promotional support by suppliers. The multiples are not in principle against welfare friendly products and do carry limited ranges of both implicit and explicit welfare friendly products but until the demand grows for these products the multiples will not engage in actively promoting them (McIntyre and Cowan, 2007).

Animal welfare and structural variations among animal supply chains

When trying to understand the dynamic of welfare initiatives in the market, it is important to recognize structural variations not only across countries and regions, but also between production sectors. Many differences between countries can actually be explained by differences between (sub)sectors and production systems, and market organization. A general feature in several countries examined here is the clear dualism between, on the one hand, large-scale, industrialized production units and, on the other, small farms with less intensive forms of production. Vertical integration is increasing everywhere, but considerable parts of the agricultural provisioning chains are characterized by fragmented structures, particularly in Italy, France and Hungary.

In France, this is first of all reflected in differences between sectors, between the beef, dairy, pig and poultry husbandry sectors. Beef farms are numerous in France (over 200,000 full-time farms) yet the average herd size is relatively modest (around 40 animals); the vast majority are pasture-based (thereby managing 20 million hectares of grassland). The farming systems are themselves highly diverse and are found all over the country. Veal farms (6,000 farms), pig farms (16,000) and poultry farms (16,000) are less numerous and more intensive, yet they are often large-scale operations and often associated with medium-sized beef farms. In general producers are well supported and the modes of production are fairly homogeneous.

In Hungary, after a period of major restructuring, including the privatization of collectives and state-owned farms, the agricultural sector is still dualistic and highly fragmented. On the one hand, there is a large number of very small farms; on the other, there are a few very large farms (0.5 per cent of the farms owning more than 50 per cent of the land) owned by large private companies. The huge numbers of small farms give input to a significant, partially informal, provisioning system for food, as represented by supply to local markets and small shops as well as own production for self-consumption and informal exchange. Although the significance of this situation has been markedly reduced over the last decade, 32–72 per cent of sales for various animal-based products are still made through these alternative (traditional) routes. Overall, pigs represent the most important (measured by value and volume) sector, followed by poultry. This is reflected in considerable exports as well as in the composition of the national diet. For Hungarians, chicken and egg production is associated mainly with small-scale, traditional production. Cattle, both dairy and beef, play a less significant role and beef is produced mainly for exports. The agricultural sector is also vertically fragmented, but a process of transformation is taking place towards more integration of production, slaughtering and processing. Major restructuring in the slaughter and processing sectors is occurring, as the sector concentrates on measures to improve cost efficiency and compliance with EU regulations. As a result, animal welfare is mainly informed by concerns over cost competitiveness for both the domestic and international markets (McIntyre and Cowan, 2007).

As the Italian population migrated from the rural south to the urban north post 1950, rural culture declined significantly. This demographic shift influenced livestock production and the consumption of animal products. A strong structural transformation has taken place and Italy has since become a net importer of cattle products (live animals, meat and milk) and pig products (piglets and pork meat). As in Hungary, the result is a clear dualism in animal production, from specialized intensive farms (localized in a few regions in the north-east of the country) that earn most of their income from this activity, on the one hand, and on the other, those who keep animals mostly for self-consumption in backyards or in small plots. The consequence of this production system is that a very small number of intensive large farms supply the mass market, while most of the other products serve local self-provisioning. The huge discrepancy in number between the two production systems allows common citizens, when crossing the countryside, to come across many 'traditional' systems (small herds, flocks in the back yard, free-range), suggesting that most animal production in Italy is still carried out in the traditional way and in small family farms. This general mis-understanding is strongly supported by food adverts on products and on TV that focus on the 'traditional' way of keeping farm animals. The disconnection between marketing of animal foods and systems of production is promoting an unrealistic image of the animal farming industry in Italy where the intensity of the animal production system is almost invisible (Miele and Parisi, 2000, 2001).

The dualism between small- and large-scale production is less evident in the other countries outlined above. The United Kingdom and the Netherlands are proponents of a more neoliberal style of governance, where public issues are increasingly addressed by way of the market and increasingly prominent farm animal welfare schemes (see Enticott, Chapter 6, this volume). Legislation has been tightened in recent years, especially in the UK as a result of foot and mouth disease (FMD) and bovine spongiform encephalopathy (BSE). UK farmers especially are under pressure as a result of export restrictions and they still have to compete with large amounts of imported meat some of which is subject to lower levels of animal welfare regulation. A substantial amount of Dutch animal products are destined for export markets, including the UK, and concern for animal welfare among foreign consumers rather than Dutch consumers supports engagement in animal friendly production and 'good welfare schemes', as is the case of selling bacon in the UK. As a baseline, Dutch producers are bound to national regulation and legislation, yet many Dutch farmers are concerned about their position in European or even global markets, which may be affected by a tightening of animal welfare legislation and any future call for a 'level playing field' in the free movement of food goods (see Lee, Chapter 3, this volume). Those farmers who produce for alternative markets, for example, organic, high quality or animal friendly, are less worried about competition, and expect an increasing consumer demand for high quality products and therefore tend to engage with animal welfare issues (Bock and van Leeuwen, 2005; Kjærnes *et al.*, 2007).

In Sweden and Norway, where the law strictly regulates farm animal welfare, only a few schemes exist and agricultural production is relatively low and oriented mainly towards the domestic market. The levels of export and import of animal products are also low compared to other countries. In Sweden, the cattle sector, in which dairy farming dominates, is larger than the pig and poultry sectors and, compared to the other sectors, more small-scale production is found here. For cattle production, unlike pig and poultry farming, there is a legal demand to keep the animals outdoors and moving on pastures during the summer period (two to four months, depending on the climatic regions). Beef producers participate less often in schemes with animal welfare components, with modernization, rationalization and concentration being prominent. Provisioning is dominated by national farmers' cooperatives who organize slaughtering and processing, yet since Sweden joined the European Union in 1995 this system has been challenged by (cheaper) imports of animal products. The current focus on Swedish origin should therefore be understood with this competition as a backdrop; Swedish food retailing is similarly dominated by three major organizations that together control around 95 per cent of the market (Roe and Marsden, 2006).

Compared to other European countries, Norwegian farming is again characterized by small units, usually family-run. This is owing to political aims of maintaining rural communities throughout the country, sustained by strong public protection of the agricultural industry. Structural changes have taken place in Norwegian husbandry over the last thirty years, but owing to regulation of, among other things, farm size, these changes have been less pronounced than in other countries. The food system is characterized by high concentration at the level of processing. Farmer-owned cooperatives for milk, meat and poultry/eggs effectively control the markets for these products. Recently, this was enhanced even more, with a merging of the meat and poultry cooperatives, which now hold more than 50 per cent of the retail market for pork and beef (and 70 per cent for slaughtering); their market share for eggs is 78 per cent. Regarding dairy products, Tine has a similarly strong position, with 93 per cent of the fresh milk market. The industry is strongly oriented towards standardized production at the level of primary production and processing, strengthened also by the logistic structure. There is a tendency now to differentiate somewhat along 'vertical', gustatory dimensions, but 'horizontal' dimensions, ranking producers, are strongly resisted (Roe and Marsden, 2006).

From these comparative national profiles it is evident that there are both common tendencies across Europe, such as intensification and vertical integration, as well as considerable variation. Some of this variation can be associated with different public agendas or 'national cultures', and with the diversity of opinion and concerns within the political and institutional contexts in which different approaches emerge. Norway and Sweden provide a specific context as law strictly regulates animal welfare and only a few private animal welfare schemes exist; with just a handful of exceptions, schemes are not reflected in differentiating product labels. And there is overwhelming consumer support for the view that farm animals are treated well (enough) and better than in other countries. Thus animal welfare emerges primarily as an issue legitimizing current state-based paternalistic practices and structures based, still (despite being in the EU internal market), upon a protectionist policy. In Scandinavia, the concentrated processing industry, cooperatively owned by farmers, takes a leading position, not so much focusing on differentiating schemes, but rather on *horizontal initiatives* such as farmers' training.

Retailer-led initiatives introducing welfare assurance and labelling schemes are found in several of the big countries, but they are only of some significance in the UK and the Netherlands, where retailer power and integration are most accentuated, and supported by legislation on 'due diligence'. What is clear here in comparing Scandinavia, the UK and the Netherlands is that it is not the level of corporate concentration that necessarily affects the style of animal welfare policy, but rather the degree to which public and state reliance and legitimacy is placed upon either public or private authorities. In France and Italy in particular, animal welfare is first of all being associated with producer-led schemes with an encompassing concept of taste and quality of various regions.

Regulating animal welfare by way of the market offers some farmers the opportunity to differentiate their products, but this may also involve conflicts of interest. British producers distrust the influence of retailers, whom they see as operating double standards; insisting on high animal welfare standards for the UK producers while also importing products of lower standards and cheaper price without communicating this difference to consumers (there are some exceptions, as described by Bowman *et al.*, 2012). Farmers also see their commercial viability being threatened by downward price pressures, while retailer margins increase. Producers want a '*level playing field*' in terms of standards with other EU countries and they also want to reap some of the added value of higher welfare standards; as is indeed the case in 'added value' chains around organic production and retailer-specific producer groups. Market strategies can be successful in improving farm animal welfare conditions in some parts of animal production, although their effect is limited within the context of globalized and liberalized markets, and there are potential 'escape routes' from animal-friendly production and consumption owing to the increasing traffic of food products by import and export across the EU (see Lee, Chapter 3, this volume).

The current Hungarian production system, where farming units for pigs and poultry dominate, supplying consumers either directly or through local markets with animal products, is regarded (by consumers) as welfare friendly, wholesome and trustworthy. Yet introducing differentiating schemes and animal welfare labelling into this somewhat unregulated trading system may be problematic and it is difficult to believe that these producers or vendors would be motivated to comply with the undoubted extra bureaucracy involved in producing to a specified welfare standard. Consumers reconcile their animal welfare concerns with the belief that the products they are consuming are produced on small-scale farms that have good welfare practices. Any shift from local markets and direct sales to big supermarkets may require new types of formalized assurances that products have reached an acceptable welfare friendly status.

There are also distinctively different patterns of retailing across Europe (Roe and Marsden, 2007). In the Scandinavian countries, a few discount supermarket chains dominate the market. In Southern and Eastern Europe, a considerable number of consumers shop for meat in independent butchers' shops and local food markets. Yet, in the United Kingdom and in the Netherlands, powerful supermarket chains offer a hugely differentiated and hierarchical assortment of quality as well as low-price goods. Finally, in countries such as France (and Germany), there is fierce competition between these different forms of distribution and retailing. Product labels are in most cases used only for pre-packaged products. A challenge for market-based strategies for improving animal welfare seems to be a lack of public control and the fuzzy line between consumer information and product marketing. Market initiatives can only be successful when consumers trust the information that is given, the foundations of which depend upon which issues are emphasized, forms of communication, and reputation. But trustworthiness will often require effective checks and independent control systems, especially if questions are being asked and trust has to be argued more actively (Kjærnes *et al.*, 2007).

The significance, contents and organization of the existing 'welfare friendly' schemes are very variable and differentially packaged (Roe and Murdoch, 2006). Most often, they take the same approach of the existing (basic EU) regulation of animal farming and refer to conditions and management systems (e.g. they describe the resources made available to the animals as in the barn systems, free-range, organic) rather than the 'actual welfare' as experienced by the animals. However, in many cases the claims on products do not refer to the actual system (sustainable or otherwise) of production (e.g. crates, cages, barns, etc.). Rather they reflect and focus upon the affective state of the animals themselves (e.g. *happiness*, *freedom*). This market mechanism has emerged within a market context in which 'claims of better animal welfare' are not required to comply with a common standard and they are made according to highly diverse criteria (Blokhuis *et al.*, 2013; Miele and Lever, 2013). It is therefore quite uncertain whether they actually contribute to the adoption of farming practices or farming innovation that improve the conditions of life for farm animals. Blokhuis *et al.* (2013) have argued that in Europe the term 'animal welfare' is interpreted in a multitude of ways and there are significant national differences in terms of how it is prioritized at country level and how economic actors are promoting market initiatives for improving it. The same authors have argued that the market approach to improving animal welfare requires a common animal welfare standard, a set of common/agreed criteria for assessing the level of animal welfare achieved in different contexts of production, different production schemes, to measure the welfare achieved by animals at farm or slaughterhouse level and to allow the comparability of products on the market. Here there are significant complications arising from the fact that 'animal welfare' is a multidimensional concept and there is no agreement in animal welfare science on which aspects of animal lives are more important or how to improve them.[3] However, a significant step in the development of a European Animal Welfare standard has emerged with the Welfare Quality Protocols (Welfare Quality, 2009a, b, c) for three species (poultry, pigs and cattle), and the EU Animal Welfare Strategy (2012–2016) has indicated the intention of the European Commission (2012) to implement these protocols for harmonizing the 'animal welfare claims' on animal products in Europe and thus increasing the animal foods' market transparency. So the EU, having recognized the increasing significance of the 'market differentiation' approach, is now attempting to provide more coherence to this in its development of generic protocols.

Conclusions: emerging marketization of animal welfare and political consumption

Across the EU there are varying degrees of institutional change taking place and varying degrees of success in attempts to improve animal welfare via labelling policies and market initiatives. The adoption of this new mechanism is increasingly advocated by the European Union as stated in the Animal

Welfare Strategy (2012–2016), and it is strongly promoted by a number of influential NGOs and corporate retailers. Partly parallel with market-based initiatives, there are tendencies to mobilize animal welfare issues in political arenas. For example, animal welfare is currently becoming a highly politicized issue in the Netherlands, such that no politician, prominent retailer or manufacturer can afford not to be interested and engaged in animal welfare issues. Rather than being interpreted as the end of the neoliberal market approach, this politicization may indeed reactivate not only public, but even private regulations.

As Koos (2012) has argued, most sociological explanations of political consumerism focus on individual consumers' motivations and resources. They address the emergence of post-materialist values, environmental or ethical attitudes, but also more traditional explanatory variables such as socio-economic characteristics, namely, income, education and gender, as important dimensions that drive political consumption. Yet the experience of the significant differences in sales of organic and 'fair trade' products, as well as in other types of politicization of the market (boycotts and *buycotts*), challenges this approach based on individual action. A growing literature points towards the importance of contextual, supra-individual factors such as globalization and economic structures in affecting shopping practices (Stolle *et al.*, 2005; Terragni and Kjærnes, 2005; Shove, 2010; Kjærnes, 2012). Since the socio-economic context outlines opportunities, constraints and incentives for action in the marketplace, the embeddedness of consumption can be expected to have a crucial impact on individual decisions to act 'politically' on the market for improving animal welfare (Evans and Miele, 2012; Koos, 2012, p. 57).

From the comparative cases presented here we see that the experiences and expectations of European consumers about the animal 'friendliness' of animal products available on the market vary and are shaped by what markets they have access to, or what the market offers them (Miele and Evans, 2010; Evans and Miele, 2012). This is in turn influenced by the interests and relative market and political power of farmers, industry and food retailers. Across Europe, consumers interpret the concept of animal welfare in various ways, but there are some common elements associated with local, small-scale and alternative production systems (Evans and Miele, 2012). These representations of animal friendliness are informing many consumption practices such as buying animal products of domestic origin, meat from production sectors considered less intensive (such as lamb), making purchases from dedicated shops and supply chains, or buying products with labels directly or indirectly indicating animal friendliness (organic, high quality, from a special region, etc.). Not all of these practices are actually contributing to improving animal welfare (i.e. buying domestic, local or typical products does not necessarily promote better animal welfare). This indicates that the market mechanism needs some specific conditions for actually promoting animal welfare, such as increased transparency, increased presence of animal friendly labelled products, increased consumer information, lower relative prices and a significant share of the population with enough disposable income to act upon political or ethical values and preferences (Koos, 2012, p. 57).

Despite this complexity, and putting to one side the increasingly controversial issue of religious slaughter (Lever and Miele, 2012), some common elements seem to emerge in the understanding of animal welfare in Europe that would align animal welfare more closely to Buller and Morris' (2009) concerns with agricultural sustainability. These include a focus on animals' 'positive emotions' such as 'happiness' (Miele, 2011); more 'natural' systems of production (free-range and organic); the presence of specific resources (space, access to pasture, high quality feed); and limits to feed additives (antibiotics, growth promoters), as well as environmentally compatible farming practices (Miele and Evans, 2010; Evans and Miele, 2012; Buller and Roe, 2013). Miele and Evans (2010) and Miele and Lever (2013) have argued that the appearance of a growing number of 'welfare friendly labelled' products have contributed to the emergence of this more situated interpretation of animal welfare in Europe, where the increasing presence of claims about the life of animals on animal products, linked to the environment, is contributing to the making of the 'animal friendly consumer' and the creation of a new modality of ethical consumption. This 'marketization' (Çalışkan and Callon, 2009, 2010; Buller and Roe, 2013) of animal welfare has also helped to establish a new norm among European citizens, that animal welfare is becoming not only a more widely shared value, a 'public good' part of a broader concept of 'sustainable agriculture' that should be sustained by specific financial incentives (e.g. in the rural development plan, cross compliance, etc.), but also as a new consumer issue deployed in the marketplace. Locating animal welfare issues in the market has significant effects. Indeed the comparative cases presented here suggest that the marketization of farm animal welfare (e.g. the emergence of a market for animal friendly products) is affecting animal welfare definitions and interpretations of animal welfare, which is increasingly becoming an 'economic' concern rather than a purely ethical concern. By this we mean that once animal welfare becomes a 'product's attribute' it also becomes subjected to processes of qualification (Callon *et al.*, 2002; Miele, 2011; Buller and Roe, 2013) by a range of economic actors, namely, the protagonists of the animal foods supply chain, to make it amenable to economic transactions. The marketization of animal welfare also requires specific devices (labels, standards, assurance schemes, etc.) and mechanisms of market 'creation' (communication and food advert campaigns) and 'market maintenance' for 'making' the new 'animal friendly consumers' (Cochoy, 2008; Miele and Lever, 2013).

Here we see that the European Union Animal Welfare Strategy (2012–2016) interprets this marketization of animal welfare as an opportunity for a common definition of animal welfare to emerge in Europe: one centred on science *and* based on common criteria for making 'animal welfare' claims on products comparable,[4] which would increase the transparency and flow of the animal products market in Europe. This mechanism will, potentially, have stronger 'harmonizing' power than the current European regulation on animal welfare which, so far, deals with the definition of a minimum standard based on a definition of minimum resources that should be made available to farm animals, but it does not indicate what level of 'welfare' animals should achieve. However, the real challenge for this mechanism to succeed is the establishment and development of a market for 'animal friendly' products in all European countries, which, as we have seen in this analysis, is currently very uneven. Although consumer concern for animal welfare is very high in all European countries (European Commission, 2007) the deployment of this concern in the marketplace is country-specific. While the most advanced examples are evident in the UK and the Netherlands, Scandinavian citizens expect all their national products to be welfare friendly and they therefore resist this model in defence of the public good nature of 'animal welfare'.

From our analysis it emerged that the farming and retailing sectors' investment in innovation and the perception of animal welfare as a competitive issue are key to the emergence of these markets and to the conditions for this orientation of the farming and retailing industry. This is consistent with Koos' (2012) general interpretation of the conditions for the emergence of political consumerism, which depends on the size of the middle class and the level of households' income in each country as well as on the presence of 'ethical complex' (Freidberg, 2004) evolving around animal welfare issues.[5] With these conditions in place, 'animal friendly consumers' willing and equipped to interpret labels in line with their knowledge about animal farming practices, and species-specific animal welfare issues (see Miele and Evans, 2010), will emerge as a product of rather than simply a driver of the market for animal friendly foods.

Notes

1 With specific reference to the UK context.
2 For ISO environmental standard please see: http://www.iso.org/iso/home/standards/management-standards/iso14000.htm.
3 As Veissier and Forkman (2008) have underlined, given the nature of the concept of animal welfare, the many dimensions that contribute to its definition and the different relevance that each aspect of animals' lives might be given by the public (naturalness, longevity, health, productivity or the emotional states … and the trade-offs among them), it is difficult to find a commonly agreed definition and, even more importantly, a preferred course of action for improving animal welfare. However, in animal welfare science there has been a significant body of research dedicated to developing indicators of specific aspects of animals' quality of life: e.g. measures for health, thermal comfort, fear, stress, pain as well as, more recently, measures for positive emotions, that could be used to assess these aspects of animals' lives in different systems of production and could be used to validate some of the claims commonly found on animal products.
4 These indicators would be based on scientifically validated measures or indicators of animal welfare, as developed, for example, in the Welfare Quality® protocols (Welfare Quality®, 2009a, b and c).
5 Susanne Freidberg (2004) defines the 'ethical complex' as a constellation of influential animal welfare and media organizations working together to highlight animal welfare issues.

References

Blokhuis, H., Jones, B., Geers, R., Miele, M. and Veissier, I. (2003) 'Measuring and monitoring animal welfare: Transparency in the product quality chain', *Animal Welfare*, vol. 12, pp. 445–455.

Blokhuis, H., Miele, M., Veissier, I. and Jones, B. (eds) (2013) *Improving Farm Animal Welfare Science and Society Working Together: The welfare quality approach*, Wageningen: Wageningen Academic Publishers.

Bock, B. and Leeuwen, F. van (2005) 'Socio-political and market developments of animal welfare schemes', in J. Roex and M. Miele (eds) *Farm Animal Welfare Concerns: Consumers, retailers and producers*, Welfare Quality Reports Series no. 1, Cardiff: Cardiff University Press, pp. 115–149.

Bock, B. and van Huik, M.M. (2007) 'Pig farmers and animal welfare: A study of beliefs, attitudes and behaviour of pig producers across Europe', in U. Kjærnes, M. Miele and J. Roex (eds) *Attitudes of Consumers, Retailers and Producers to Farm Animal Welfare*, Welfare Quality Report Series no. 2, Cardiff: Cardiff University Press, pp. 73–121.

Bowman, A., Froud, J., Johal, S., Law, J., Leaver, A. and Williams, K. (2012) 'Bringing home the bacon: From trader mentalities to industrial policy', CRESC webpage, available at: www.cresc.ac.uk/publications/bringing-home-the-bacon-from-trader-mentalities-to-industrial-policy, accessed 10 March 2013.

Buller, H. and Cesar, M. (2007) 'Eating well, eating fare: Farm animal welfare in France', *International Journal of Sociology of Food and Agriculture*, vol. 15, no. 3, pp. 45–58.

Buller, H.J. and Morris, C. (2009) 'Beasts of a different burden: Agricultural sustainability and farm animals', in S. Seymour, R. Fish and S. Watkins (eds) *Transdisciplinary Perspectives on Sustainable Farmland Management*, Wallingford: CABI, pp. 135–148.

Buller, H. and Roe, E.J. (2011) *Certifying Quality, Negotiating and Integrating Animal Welfare into Food Assurance*, in Welfare Quality Reports Series no. 15, Cardiff: Cardiff University Press.

Buller, H.J. and Roe, E. (2013) 'Modifying and commodifying farm animal welfare: The economisation of layer chickens', *Journal of Rural Studies*. Available at: http://www.sciencedirect.com/science/article/pii/S0743016713000077.

Çalışkan, K. and Callon, M. (2009) 'Economization, part 1: Shifting attention from the economy towards processes of economization', *Economy and Society*, vol. 38, no. 3, pp. 369–398.

Çalışkan, K. and Callon, M. (2010) 'Economization, part 2: A research programme for the study of markets', *Economy and Society*, vol. 39, no. 1, pp. 1–32.

Callon, M., Meadel, C. and Rabeharisoa, V. (2002) 'The economy of qualities', *Economy and Society*, vol. 31, pp. 194–217.

Cochoy, F. (2008) 'Calculation, qualculation, calqulation: Shopping cart arithmetic, equipped cognition and the clustered consumer', *Marketing Theory*, vol. 8, no. 1, pp. 15–44. DOI: 10.1177/1470593107086483.

European Commission (2007) 'Attitudes of EU citizen towards animal welfare: Special Eurobarometer 270', Wave tt.1, TNS Opinion 8: Social, Brussels: European Commission.

European Commission (2012) 'Communication from the Commission to the European Parliament, the Council and the European Economic and Social Committee on the European Union Strategy for the protection and welfare of animals 2012–2015' (Text with EEA relevance), {SEC(2012) 55 final}, {SEC(2012) 56 final}, 15.2.2012 COM(2012) 6 final/2, Brussels: European Commission.

European Union (1997) Treaty of Amsterdam, Luxembourg: Office for Official Publications of the European Communities.

Evans, A. and Miele, M. (2012) 'Between flesh and food: How animals are made to matter or not to matter in food consumption practices', *Environment and Planning D – Society and Space*, vol. 30, no. 2, pp. 298–314.

Fodor, E. and Redai, D. (2006) 'National report Hungary, Socio-economic trends and welfare policies', Deliverable 3.1 VI Framework EU project Quality – November 2006, available at: https://www.jyu.fi/ytk/laitokset/yfi/oppiaineet/ykp/tutkimus/paattyneet/laatu/D3.1%20Combined%20National%20Reports%20-%20Final.pdf, accessed 10 October 2013.

Francione, G. and Gardner, R. (2010) *The Animal Rights Debate: Abolition or regulation?* New York: Columbia University Press.

Freidberg, S. (2004) 'The ethical complex of corporate food power', *Environment and Planning D – Society and Space*, vol. 22, no. 4, pp. 513–531.

Ingenbleek, P.T.M. and Immink, V.M. (2011) 'Consumer decision-making for animal friendly products: Synthesis and implications', *Animal Welfare*, vol. 20, pp. 11–19.

Kjærnes, U. (2012) 'Ethics and action: A relational perspective on consumer choice in the European politics of food', *Journal of Agricultural and Environmental Ethics*, vol. 25, no. 2, pp. 145–162.

Kjærnes, U. and Lavik, R. (2007) 'Farm animal welfare and food consumption practices: Results from surveys in seven countries', in U. Kjaernes, M. Miele and J. Roex (eds) *Attitudes of Consumers, Retailers and Producers to Farm Animal Welfare*, Welfare Quality Reports Series no. 2, Cardiff: Cardiff University Press, pp. 1–29.

Kjærnes, U., Bock, B. and Miele, M. (2009) 'Improving farm animal welfare across Europe: Current initiatives and venues for future strategies', in U. Kjarnes, B. Bock, M. Higgin and J. Roex (eds) *Farm Animal Welfare within the Supply Chain: Regulation, agriculture and geography*, Welfare Quality Reports Series no. 8, Cardiff: Cardiff University Press, pp. 1–69.

Kjærnes, U., Harvey, M. and Warde, A. (2007) *Trust in Food: A comparative and institutional analysis*, Basingstoke: Palgrave Macmillan.

Koos, S. (2012) 'What drives political consumption in Europe? A multi-level analysis on individual characteristics, opportunity structures and globalization', *Acta Sociologica*, vol. 55, no. 1, pp. 37–57.

Lever, J. and Miele, M. (2012) 'The growth of the halal meat markets in Europe: An exploration of the supply side theory of religion', *Journal of Rural Studies*, vol. 28, no. 4, pp. 528–537.

McIntyre, B. and Cowan, C. (2007) 'Hungarian consumers' views about animal welfare', in A. Evans and M. Miele (eds) *Consumers' Views about Farm Animal Welfare Part I: National reports based on focus group research*, Welfare Quality Reports Series no. 4, Cardiff: Cardiff University Press, pp. 1–52.

Marres, N. (2009) 'Testing powers of engagement: Green living experiments, the ontological turn and the undoability of involvement', *European Journal of Social Theory*, vol. 12, pp. 117–133.

Mayfield, L., Bennett, R. and Tranter, R. (2007) 'United Kingdom consumers' views about animal welfare', in A. Evans and M. Miele (eds) *Consumers' Views about Farm Animal Welfare Part I: National reports based on focus group research*, Welfare Quality Reports Series no. 4, Cardiff: Cardiff University Press, pp. 115–154.

Miele, M. (2011) 'The taste of happiness: Free-range chicken', *Environment and Planning A*, vol. 43, no. 9, pp. 2076–2090.

Miele, M. and Evans, A. (2010) 'When foods become animals: Ruminations on ethics and responsibility in care-full spaces of consumption', *Ethics, Policy and Environment*, vol. 13, no. 2, pp. 171–190.

Miele, M. and Lever, J. (2013) 'Civilizing the market for welfare friendly products? The techno-ethic of the Welfare Quality© assessment', *Geoforum*, vol. 48, pp. 63–72.

Miele, M. and Parisi, V. (eds) (2000) *Atteggiamento dei consumatori e politiche di qualità della carne in Italia e in Europa negli anni novanta*, Milan: Franco Angeli.

Miele, M. and Parisi, V. (2001) 'L'Etica del Mangiare, i valori e le preoccupazioni dei consumatori per il benessere animale negli allevamenti: un'applicazione dell'analisi Means–end Chain', *Rivista di Economia Agraria*, Anno LVI, no. 1, pp. 81–103.

Miele, M., Ara, A. and Pinducciu, D. (2007) 'Italian consumers' views about animal welfare', in A. Evans and M. Miele (eds) *Consumers' Views about Farm Animal Welfare Part I: National reports based on focus group research*, Welfare Quality Reports Series no. 4, Cardiff: Cardiff University Press, pp. 53–115.

Miele, M., Murdoch, J. and Roe, E. (2005) 'Animals and ambivalence: Governing farm animal welfare in the European food sector', in V. Higgins and G. Lawrence (eds) *Agricultural Governance*, London: Routledge, pp. 169–185.

Miele, M., Blokhuis, H., Bennett, R. and Bock, B. (2013) 'Changes in farming and in stakeholder concern for animal welfare', in H. Blokhuis, M. Miele, I. Veissier and B. Jones (eds) *Improving Farm Animal Welfare. Science and Society Working Together: The Welfare Quality© approach*, Wageningen (NL): Wageningen Academic Publisher, pp. 19–47.

Murdoch, J. (2005) 'A comparative analysis of retail structures across six countries', in J. Roex and M. Miele (eds) *Farm Animal Welfare Concerns: Consumers, retailers and producers*, Welfare Quality Reports Series no. 1, Cardiff: Cardiff University Press, pp. 83–111.

Petterson, L. and Bergman, H. (2007) 'Swedish consumers' views about animal welfare', in A. Evans and M. Miele (eds) *Consumers' Views about Farm Animal Welfare Part I: National reports based on focus group research*, Welfare Quality Report Series no. 4, Cardiff: Cardiff University Press, pp. 205–252.

Poulain, J. P., Tibère, L. and Dupuy, A. (2007) 'French consumers' views about animal welfare' in A. Evans and M. Miele (eds) *Consumers' Views about Farm Animal Welfare Part I: National reports based on focus group research*, Welfare Quality Reports Series no. 4, Cardiff: Cardiff University Press, pp. 323–371.

Roe, E. and Marsden, T. (2007) 'Analysis of the retail survey of products that carry welfare-claims and of non-retailer led assurance schemes whose logos accompany welfare-claims', in U. Kjaernes, M. Miele and J. Roex (eds) *Attitudes of Consumers, Retailers and Producers to Farm Animal Welfare*, Welfare Quality Reports Series no. 2, Cardiff: Cardiff University Press, pp. 33–65.

Roe, E. and Murdoch, J. (2006) *UK Market for Animal Welfare Friendly Products: Market Structure, Survey of Available Products and Quality Assurance Schemes*, Welfare Quality Reports Series no. 3, Cardiff: Cardiff University Press.

Schipper, L., Beekman, V. and Korthals, M. (2007) 'Netherlands consumers' views about animal welfare', in A. Evans and M. Miele (eds) *Consumers' Views about Farm Animal Welfare Part I: National reports based on focus group research*, Welfare Quality Report Series no. 4, Cardiff: Cardiff University Press, pp. 155–204.

Shove, E. (2010) 'Beyond the ABC: Climate change policy and theories of social change', *Environment and Planning A*, vol. 42, no. 6, pp. 1273–1285.

Stolle, D., Hooghe, M. and Micheletti, M. (2005) 'Politics in the supermarket: Political consumerism as a form of political participation', *International Political Science Review*, vol. 26, no. 3, pp. 245–269.

Terragni, L. and Kjærnes, U. (2005) 'Ethical Consumption in Norway: Why is it so low?', in M. Bøstrom (eds) *Political Consumerism: Its motivations, power, and conditions in the Nordic countries and elsewhere*, TemaNord 2005:517, Copenhagen: The Nordic Council of Ministers.

Terragni, L. and Turjusen, H. (2007) 'Norwegian consumers' views about animal welfare', in A. Evans and M. Miele (eds) *Consumers' Views about Farm Animal Welfare Part I: National reports based on focus group research*, Welfare Quality Reports Series no. 4, Cardiff: Cardiff University Press, pp. 253–322.

Veissier, I. and Forkman, B. (2008) 'The nature of animal welfare science', *ARBS Annual Review of Biomedical Science*, vol. 10, pp. 15–26.

Welfare Quality® (2009a) *Assessment Protocol for Cattle*, Welfare Quality® consortium, The Netherlands: Lelystad.

Welfare Quality® (2009b) *Assessment Protocol for Poultry*, Welfare Quality® consortium, The Netherlands: Lelystad.

Welfare Quality® (2009c) *Assessment Protocol for Pigs*, Welfare Quality® consortium, The Netherlands: Lelystad.

8 Exploring the new rural–urban interface

Community food practice, land access and farmer entrepreneurialism

Alex Franklin and Selyf Morgan

Introduction

Studies of rural–urban linkages have become prominent in attempts to understand processes of agricultural restructuring, reflecting an increasingly complex and dynamic relationship between rural and urban spaces. At the same time, knowledge and awareness of community-led sustainability action contributes directly to understanding the steady increase in spaces of agricultural practice within urban settings; in particular, through an ever growing number of community food producing initiatives. A wide range of factors inform and sustain these initiatives. In addition to food system-specific drivers, local food initiatives[1] are often perceived as effective vehicles for encouraging residents to increase local community engagement. Accordingly a wide range of social and environmental, as well as economic reasons are regularly cited in support of the argument that local food initiatives make positive contributions – both tangible and intangible – to community resilience.

Beyond their role in increasing community cohesion and sustainability practice in their neighbourhoods, local food initiatives have the potential to support sustainable place-making at a city-region level by strengthening connections between urban and rural communities. However, taking the UK as a reference point, there currently appears to be considerable variation in the extent to which urban-based local food initiatives are actively supported and engaged with by members of traditional farming communities. In this chapter we look first to identify some of the reasons for this situation, and second to consider ways that farmers may be encouraged to collaborate and engage with a wider range of local food initiatives. In addition, we also aim to contribute to current understandings of rural restructuring and urban–rural linkages more broadly. We do so by considering community food initiatives within the wider context of different forms of multifunctional, multi-user, community land-based practices (referred to here as 'community landshare' initiatives) in a range of urban, peri-urban and rural settings. Specifically, we are interested in the connections that exist between community landshare and the individual actions of land owners.

There is a noticeable contrast between the rich variety of experimentation and innovation found among urban-based community food initiatives, and the more limited extent to which farmers and other rural landowners have been able to participate in direct and interactive relationships with community groups in all settings. While we review instances where such interaction has been developed we suggest that current agri-food, property rights and land-use paradigms have maintained barriers that are inhibiting more intimate and creative relationships from taking root. Also pivotal here is the role of social relationships within community-led local food initiatives. This includes farmer involvement in practices such as urban agriculture; farmers' markets; the establishment of alternative food networks and supply chains (e.g. food hubs) as well as various other ways in which community food groups and farmers are responding to growing demand among the urban populous for more 'meaningful' engagement with greenspace and the natural environment. We are, therefore, also led to consider farmer entrepreneurialism and how changes in the EU Common Agriculture Policy (CAP) encourage entrepreneurial behaviour. In doing so we employ a simple typology to identify different expressions of entrepreneurial behaviour among farmers and how they may affect their interaction with community groups.

The questions that we use to frame the broader discussion of urban–rural linkages and community-led action include: how can community interest in food growing practices in urban areas be better integrated with farm practice in rural areas; how do differences in the needs of community members and farmers shape the nature and geography of these linkages and practices, and; what are the implications thereof for rural policy and practice? By way of context, we begin by reviewing the contribution of European agricultural and rural policy to supporting the development of multifunctional land-based practices in rural and peri-urban settings, and in turn, how this can lead to different forms of entrepreneurial behaviour among farmers.

Rural re-structuring: multi-functionality and farmer entrepreneurialism

Over the last twenty to thirty years the demands on European farmers have changed and been extended in a number of different directions. Agricultural reform since the late 1980s has been accompanied by a widespread reappraisal of rural space and a reduction in the central role of agriculture to the extent that rural Europe is now less defined by the production capacities of farming than it was in the past (see Lee, Chapter 3, this volume). A large literature discussing these changes has accumulated, covering changes in policy, changes in farmer behaviour, and the evolution of the wider rural economy. The following discussion is not a comprehensive review of these issues, but indicates some of the main developments that provide opportunities for greater integration between farmers and enhance the involvement of different communities in food production.

Change in the role of agriculture has been underpinned by a shift in policy instruments away from price support and production-based payments (Pillar I measures in CAP) towards developing the role of agriculture beyond the production of food and fibre. Changes to CAP during the 1990s led to the creation of the Rural Development Regulation (RDR) and the so-called Pillar II of CAP. These changes and the RDR in particular, connected rural development more directly with agricultural policy through the creation of Rural Development Plans (RDPs), which are designed in part to encompass the multifunctional contributions of agriculture to the rural economy (e.g. Dwyer *et al.*, 2007).

The mid-term (Fischler) review of the CAP reform process in 2003 brought further change to Pillar I measures, and enhancements to Pillar II. Major alterations to Pillar I saw the introduction of the Single Payment Schemes (SPS), which has largely removed the direct linkage between support payment and production. Agricultural support was further regulated by the introduction of cross-compliance conditions, to ensure that land is maintained in 'good agricultural order' (EC, 2003), further reflecting the shift of focus that underlay CAP reform. As well as promoting multifunctionality through Pillar II support for activities that ameliorate the negative environmental and social impacts of agriculture, the cross-compliance conditions also provide support for rural enterprises that may not be directly associated with farms and farming (EC, 2004).

Largely owing to these changes, the 1990s and early 2000s in Europe are regarded as a period when a post-productivist approach to rural and agricultural policy gained prominence, and when the traditional policy of agricultural exceptionalism diminished (see Lee, Chapter 3, this volume). Rural areas were increasingly regarded as spaces of consumption as well as production, providing environmental and amenity goods as well as agricultural commodities (e.g. Marsden *et al.*, 1993). Issues of farm animal health and welfare, and food safety became much more prominent during this period, while broader concerns, such as climate change, also began to make substantial inroads into thinking about food production systems. Agricultural, food and rural policy domains thus expanded to include new values and concerns driven by actors who were different to those previously prominent (e.g. Daugbjerg and Swinback, 2012). This shift was not a purely European phenomenon, and it can be argued that agricultural policy reform has been encouraged by greater environmental concern, diversification in production practices, and a wide range of rural development measures undertaken in a number of countries across the world (Diakosavvas, 2006; Grant, 2012).

Ideas about the multifunctional nature of agriculture have accompanied and contributed to agricultural policy reform, and there has been a lively debate, extending well over a decade, about what agricultural multifunctionality might entail (for a review see Wilson, 2007). The OECD (2001) provided an early definition, distinguishing two forms: on the one hand, multifunctionality as an inherent characteristic of agriculture, where the joint production of commodity and non-commodity output is recognised; and on the other, as a normative concept that charges agriculture with objectives that include social and other valued goals. Both forms may coexist and be present in differing combinations and strength (Wilson, 2008) and can also be considered in relation to a wider discussion about different forms or paradigms of agriculture. Marsden (2003), for example, describes three distinct models at the European scale. These three models are labelled as the 'agri-industrial', the 'post-productivist' and the 'rural development' models, all three of which may be observed throughout the EU.

The description of the multifunctional nature of agriculture and of different forms of agriculture, is interwoven with the political, social and economic drivers of policy reform, producing complex descriptions of agriculture and, by implication, farmers. We can thus expect that farmers operating within the three models proposed by Marsden (2003) are likely to present different identities, expectations and styles of farming. The *agri-industrial* model, associated with the globalised production of standardised products, faith in free competition and in the application of technology intensive solutions that reduce costs through economies of scale, suggests a view of farmers as production maximisers. In the *post-productivist* model, the agricultural sector is small in relation to other economic sectors and is decreasing in economic relevance. In this model the rural landscape is a consumption-good for the urban population; agriculture, while continuing to be productive, is no longer central but regarded as providing services and a context to other land uses. Farmers in this model adopt various forms of land management regimes, not all of which are geared to food production. The third model is focused on the concept of *sustainable rural development*, which is seen as a set of practices that has the potential to reconnect productive agriculture with people within rural and urban areas (see Sonnino *et al*., 2008). Agriculture in this model is intimately combined with the socio-economic health of rural areas and recognised as an economic sector that must be integrated into the wider economy as part of a sustainable rural development paradigm (Marsden and Sonnino, 2008). Farmers operating under these latter principles differ from those of the other two models in being open to new and more intimate relationships with communities that share an interest in, and aim for more involvement with, food production.

A caveat to this narrative of multiple and to some extent competing models of agriculture should, however, be acknowledged. Its robustness has been challenged in the latter half of the 2000s as the trends that contributed towards weakening the productivist model of agriculture have been checked. A range of macro-scale issues, including a nutrition transition effect as a wealthier global population consume more meat; a forecast global population of some nine billion by 2050; increased competition for land, water and energy; and the unpredictable effects of climate change, have led to increased concerns about food security in its various forms (see Morley *et al*., Chapter 2, this volume). World commodity prices have been rising, and in lieu of price spikes in the mid 2000s, the EU suspended the agri-environmental set-aside requirements and proceeded to abolish them in the Health Check review of 2007/2008. A resurgence of a productivist mentality, supply control, and renewed faith in the power of science and technology to provide technical fixes, such as represented by GM technologies and the 'sustainable intensification' approach, can be identified. In parallel, the various processes of globalisation, exemplified not least by the WTO trade talks have added to uncertainties concerning the direction of policy travel and further change to the CAP. The European Commission consultation paper on CAP reform post-2013 (EC, 2010), for instance, recommended that the strategic aims of the CAP should be to preserve food production potential on a sustainable basis throughout the EU, to guarantee European food security as well as contribute to satisfying growing global demand. Accordingly, more commentators are prepared to argue that the importance of multifunctionality, which might be identified as one of the strongest themes of recent multi-paradigm narratives, has been reduced. There is also some suggestion that these concepts had always been used as devices to justify subsidy regimes, only coincidentally addressing multifunctional objectives as a policy fix (e.g. Daugbjerg and Swinback, 2011).

Whatever the ultimate significance and future of multifunctionality in policy development, the cumulative effect of policy changes to CAP has animated discourse and encouraged real action in support of the multifunctional model of agriculture. This support remains consistent with ongoing efforts to reduce the financial burden of CAP and farmer dependence on state aid (Moyer and Josling, 2002; EC, 2010) and is consistent with the need to respond to pressures for reform emanating from trade talks. Farmers thus continue to be encouraged to become more competent market actors and to develop their entrepreneurial skills and decision making. In short, entrepreneurship has become an important component of the restructuring process in European agriculture.

Entrepreneurial behaviour gives rise to different types of business strategies. These may be related to different models of agriculture, with business strategies reflecting commercial contexts pertinent to each model. Morgan *et al.* (2010) discuss three strategies which are broadly consistent with Marsden's (2003) models, namely a 'bulk' farming strategy, where the focus is on production; a 'value-adding (VA)' strategy, where the farmer adds value from downstream food-related activity; and 'non-food diversification (NFD)', where the farmer uses the farm, its land and farming activity as the basis for other business ventures. This classification focuses on entrepreneurial skills, ensuring that the designation of strategy type does not preclude farmers from employing more than one approach. Entrepreneurial skills, as defined in Morgan *et al.* (2010), include creating and evaluating the business strategy, networking and utilising contacts, and recognising and realising opportunities. Bulk production farmers tend to be committed to more rigid and structured relationships, particularly with downstream partners, with a narrower scope for entrepreneurial action compared to those farmers who add value to farm output, or are engaged in non-food diversification. Farmers employing these latter strategies are, in turn, able and required to access more diverse and fluid networks and opportunities.

Farmers who developed links with the wider (rural and urban) economy may also develop new market channels that provide alternatives to the conventional agri-food sector. Sonnino and Marsden (2006) argue that the alternative nature of these networks is associated with the way that they are created through reforming social, ecological and economic relations, and are contingent on the economic, ethical and ecological stances of participating farmers. Farmers exploit the inherent character of alternative farming systems and of the foods produced (e.g. organic farmers), while others build on past traditions (e.g. production practices, quality conventions) that are translated and represented in terms that communicate the value of those traditions to contemporary retailers and consumers. While we do not exclude from consideration those farmers whose business strategy predominantly entails maximising production (the 'bulk' farmers), it is farmers whose networks extend horizontally to interact with the wider rural and urban economies that we focus on when examining how community interest in food growing practices may be better integrated with farm practice.

Neo-productivism and urban planning

To date, as presented above, policy and academic discussions on the implications of agricultural restructuring and CAP reform have been focused predominantly on rural space and the farm-based economy. Little consideration has been given to changes in the urban realm. In this sense, policy and governance structures have also largely maintained the established post-war consensus around protection (of the rural) and containment (of the urban) embodied in the UK 1947 agriculture and town and country planning legislation. More recent discussions of multifunctional agriculture and farmer entrepreneurialism have only obliquely been related to these changing urban contexts, and are narrowly expressed through consideration of the value-added activities that farmers may undertake. Links to the urban context have been more in terms of supplier–market relationships and less in terms of a joint enterprise in restructuring these links. Increasingly, however, these limitations are being tested by urban-based actors through a number of initiatives and social innovations, some of which may be classified as 'neo-productivist'.

The term 'neo-productivism' is used here to conceptualise the growing desire shown by members of civil society to become more meaningfully and seriously engaged with greenspace and the natural environment, in either a rural or an urban setting. Arguably neo-productivism encompasses a strong social dimension. This definition draws on, but also goes beyond recent work by a number of scholars around rural restructuring and multifunctional land use. Ravenscroft and Taylor (2009), for example, refer to a transition towards a 'deeper and more sustained relationship between people and the land', realised through engagement in 'more ethical and environmentally friendly' forms of production, or 'serious leisure' (Stebbins, 1982; Cohen-Gewerc and Stebbins, 2013). Consistent with other recent academic work on neo-productivism, however, they limit their focus to a rural context. A study of community food initiatives suggests the potential for broadening the gaze of neo-productivism both to urban spaces and to whole groups of individuals. As a starting point, we provide a brief introductory overview to the concept of multifunctionality in an urban planning context and consider current policy approaches to the protection and promotion of urban greenspace. We then focus on urban community-led local food initiatives.

During the post-war years city-wide urban density largely decreased across much of the UK. Despite the introduction of a policy to maintain green belts around the larger conurbations, together with the fierce protection of agricultural land through the twin pillars of the 1947 Agriculture Act and 1947 Town and Country Planning Act, this period saw a gradual but highly regulated expansion of city boundaries. Integral to this expansion, but also a guiding feature in the parallel construction of new urban settlements (as a way of controlling urban sprawl), was the shift from multifunctional to monofunctional land use, with sectors such as housing, business, transport, agriculture and recreation becoming increasingly separated out (Priemus and Hall, 2004; Priemus *et al.*, 2004). Since approximately the 1980s, however, there has been a resurgence of 'compact city' informed development and with it, a revival of multifunctional land use planning. For some commentators the wide-ranging benefits of compact city development are perceived to make them 'theoretically more resilient in the context of climate change, with fewer harmful impacts' (UNHabitat, 2009, p. 58). However, despite the rhetoric surrounding the wholesale advantages of compact city development, the same level of support is not always found for many of the characteristics of such developments by its actual urban residents. Increasingly, urban planners are coming under pressure to provide high quality urban amenities, urban culture and urban life, but with the retention of accessible greenspace as an essential component of this, even within the heart of the city (Priemus and Hall, 2004). One of the oft cited attractions of living in suburban areas is a relatively high level of private and public greenspace when compared to urban centres. Accordingly, to secure and sustain a mixed socio-economic demographic in a city's core, imaginative ways of pairing good access to open space and compact living will become increasingly critical to future strategies for sustainable urban planning. Also relevant in this context is the growing acceptance among professional urban planners of the much broader range of economic and environmental benefits that are supported by adequate provisioning and protection of urban greenspace.

Although initiatives to promote greenspace in line with social objectives can be traced back at least as far as the nineteenth century with the creation of urban parks (Swanwick *et al.*, 2003), the wider value of greenspace relative to further development of the built environment has had a more varied history. The inclusion of greenspace (or 'green infrastructure') as part of a more strategic spatial planning of city-regions nowadays occurs in a range of different forms, as linked to different social, economic and environmental objectives. This includes greenways, green corridors, green links and green wedges (Swanwick *et al.*, 2003). At the same time, the potential is also gradually being realised for the reintroduction of nature into city centres through optimum use of green building design. Good access to greenspace is also deemed essential to promoting the health and well-being of urban residents. In many ways this awareness of access to greenspace acts as an indicator that sustainable urban living actually represents a return to previous ideals of the early twentieth century (Mehmood, 2012).

Since the 1990s, the protection of urban greenspace has also attracted considerable support as a way of adapting to climate change. This reflects a gradual, yet growing recognition of the full and interconnected functionality potentials of this resource. For example, one of the important environmental uses of greenspace is the balance it provides for the impermeable hard surfaces of 'grey' space (Swanwick *et al.*, 2003). In acknowledgement, however, not only of its ecological importance, but also its social and economic significance, Swanwick *et al.* (2003) discuss how greenspace has both an 'existence value' – 'because people know it is there' – and a 'use' value (p. 100). As this also leads Gill *et al.* (2007) to assert: 'the priority for planners and greenspace managers is to ensure that the functionality of greenspace is properly understood and that what exists is conserved' (p. 130). However, despite increasing recognition by practitioners and academics alike of the multidimensional importance of urban greenspace, as yet only very limited priority appears to have been given to planning for more 'meaningful engagement' with areas of greenspace by urban residents. Accessibility is often simply defined in terms of the ability of an individual to reach an area of public greenspace on foot (for example, a park, garden or playing field), within an acceptably short period of time. It is in this sense that the above call by Gill *et al.* (2007), for urban planners to develop a more comprehensive understanding of the full potential functionality of urban greenspace, remains largely unanswered.

Arguably, where urban policy prioritisation is limited to ensuring good accessibility to local areas of greenspace, this obscures the equally important role that public bodies and other urban stakeholders can play in supporting neo-productivist forms of public engagement within urban settings. At the same time, though, also of direct relevance here is the widespread disconnect between urban planning policy and the regulatory arena of agricultural productivism. That both pillars of the CAP remain firmly orientated towards rural space serves as a reflection of the ongoing lack of recognition given to either urban greenspace or urban-based local food initiatives in the context of sustainable food planning. Currently, it is among the third (voluntary) sector that the momentum for facilitating multifunctional neo-productivist uses of urban greenspace lies. Community-led local food growing initiatives serve as a case in point.

Meaningful engagement, community food growing initiatives and access to land

An increase in the 'meaningful engagement' of a group of individuals in a range of different natural resource-based production or consumption practices is a common feature of local food initiatives, particularly where they are managed by a core group of local community residents. Notably, though, while such initiatives support some degree of collective participation, they vary in the extent to which they enable (or require) participants to become involved in *both* the production and consumption of local food. They also vary in the extent to which they engage the services of full-time local food producers (i.e. professional farmers) and whether or not they operate across both rural and urban space. Commonly this variation depends on whether:

- an initiative is focused around supporting universal participant involvement in food growing practices through bringing together food consumers and spaces of food production, for example urban food growing spaces and community allotments;
- the primary aim is only to increase connections between the actors of production and consumption, for example farmers' markets and farm shops;
- the initiative is intentionally designed in a bid to achieve increased connections between multiple actors of production and consumption, with regular opportunities for their interaction in spaces of food production, for example community supported agriculture (CSA) and (some forms of) community food hubs.

In reviewing the contribution of community food initiatives in supporting more meaningful engagement with greenspace and strengthening urban–rural relations, this variation is potentially significant. Arguably, it suggests that there is a need for more specificity in the type of community food initiative being referred to when considering either the contributions they make, or the potential limitations of their reach. Hinrichs (2003, p. 41) for instance, observes how 'direct marketing initiatives, such as farmers' markets and CSAs, foster greater mutuality, by bringing producers and consumers together through face-to-face contact'. Hinrichs does not go on to comment, however, on how this 'mutuality' might be implicated by the marked differences that exist between these two types of initiatives in their respective retention and blurring of the socio-spatial boundaries between practices of production and consumption.

From the above categorisation of community food initiatives, it is the first and third types that appear most able to increase the opportunities for urban residents to become more meaningfully engaged with greenspace. Notably, however, a review of the literature suggests that it is among these two categories that far less progress has thus far been made in securing the parallel active engagement of the local farming community, or in incorporating rural space. This raises the question of whether meaningful engagement with greenspace in the case of these two categories is actually limited to urban areas; and if so, what are the barriers to extending opportunities for meaningful engagement beyond urban boundaries and how might they be overcome? A useful starting point for addressing these two questions is the availability of land.

The centrality of land access to both community food initiatives and other community-level sustainability initiatives makes the current lack of academic discussion of the significance of land ownership in this context somewhat surprising. As Brown (2007) asserts, quite simply 'property matters'; it plays a profound role in mediating how people engage with land. Some community groups are able to maximise their property rights by purchasing land collectively and holding it in trust. Very often, however, the market value and limited availability of accessible land (especially in urban or peri-urban areas), prohibits community buy-outs. It is also rarely an option in the case of a recently formed community group; a group whose membership does not include a sufficient number of socially empowered individuals; a group who do not have the financial resources, or a group who simply do not want the commitment and responsibility of land ownership. In these cases initial access is dependent upon the willingness of a third-party landowner. The distinction between whether land is in public, private or third sector ownership is, however, not in itself necessarily prohibitive to supporting collective forms of community food growing. Rather, what matters are the socio-legal relationships that flow between landowners and land users. In particular, the landowner should feel confident that his or her interests in an area of land will in no way be threatened as a result of allowing it to be used for a community initiative.

One good illustration of where a large number of landowners recently became engaged with supporting community food growing in an urban setting, is the London-based Capital Growth initiative. Launched in November 2008, Capital Growth is a partnership between London Food Link, the (Big Lottery-funded) Local Food programme, and the Mayor of London. One of the principal cited motivations behind the establishment of Capital Growth was the lack of sufficient provisioning of urban allotments (a problem that also extends across most of the UK). The original target set and exceeded by Capital Growth was to create 2,012 new community food growing spaces in London by the end of 2012. A new food growing space was defined as being 'community led, involving five or more unrelated people; and being at least five square metres in size' (Sustain, 2013).

In meeting the target of 2,012 spaces the partners were able to tap into the media hype around the 2012 London Olympics. In total, some 500,000 square metres of urban land had reportedly been turned to food production by the end of 2012, with approximately 99,000 people engaged (Sustain, 2013). Of the new growing spaces created, 35 per cent were located on school land, 20 per cent in housing estates and 66 per cent on previously unused, derelict or inaccessible land.

Examples such as Capital Growth support the claim that there appears to be an increase in the willingness of urban landowners to offer limited rights of land use to community groups. In granting these rights the landowners are enabling a wider cross-section of society to become more actively engaged in food production practices. Also apparent, however, from the Capital Growth example (although not explicitly discussed in the official project report), is the key role of the Capital Growth partnership as a *trusted intermediary* between landowners and potential land users. Having the involvement of a professional team who were able to 'get permission from the land owner', as was somewhat overly simplistically stated in the final report (Sustain, 2013, p. 7), is what made it possible for this initiative to reach a city-wide scale. The importance of there being a trusted intermediary involved is due to the fact that a wide range of concerns commonly exist among landowners in connection with making land available to community groups. For example, the Federation of City Farms and Community Gardens (FCFCG), which is a UK-based third sector network organization, reported that these concerns can include: the potential for subsequent development and planning delays, scepticism about the accountability and capabilities of community projects, concerns about project failure and bad publicity, fear of longer-term loss of control of the resource or of an area of land to shift from being an asset to being a liability for the owner. Other potentially relevant factors include the degree of fit with an owner's own plans and aspirations for the land, the proposed duration of community use, whether the type of use and number of individuals involved is likely to result in any damage to the land, and whether any other foreseen costs or benefits of making the land available are deemed acceptable (FCFCG, 2012). Notable from the research undertaken by the FCFCG is that many of the potential barriers to securing the involvement of landowners in community food growing initiatives can be the same regardless of whether the land is located in an urban or rural setting. However, whether or not these barriers turn out to be negotiable or insurmountable will often depend on a more micro-level set of relational issues that vary from case to case.

For an agreement to be reached it commonly requires some degree of mutual benefit or at the very least a shared interest, identity or ideological belief to exist between a landowner, the prospective community user group and their proposed use of the land. These micro level factors will also inform the degree to which a landowner is himself/herself prepared to become directly involved with facilitating collective practice. For in the case of some community food initiatives, it is not always just access to land that is required. Where the landowner is a farmer, their active involvement is often also necessary for the initiative to be a success. This is particularly so in the case of initiatives that are aimed at increasing both connections between actors of food production and consumption, and joint practice between these two sets of actors within spaces of food production; or initiatives where there is a deficit of land/animal husbandry skills and knowledge on the part of the community group. Significantly though, this requires that the farmer–landowner is willing to operate in accordance with the founding principles of the initiative. Where these principles are perceived to be overly restrictive (e.g. organic certification) or particularly alternative (e.g. bio-dynamic agriculture) to the methods usually followed by the farmer, or if they are perceived to offer limited opportunities for mutual benefit, then their active engagement can prove extremely challenging to secure. Much will depend on the degree of flexibility that either side are prepared to bring to the table. In the case of farmers, the altruistic motivations for supporting community food initiatives will also need to be balanced with the bottom-line economic costs of running a farm. The case of Stroudco community food hub (England) provides a useful illustration here, including the challenges of following a path that attempts to address all of these demands.

Community food hubs can be broadly defined as coordinating alternative sourcing, supply, and/or marketing on behalf of producers and consumers, and providing technical as well as infrastructure support for product distribution. In addition to having clear environmental goals, they are also often founded on social motivations relating to community cohesion, social gain, increasing healthy eating options, and improving local food access options. At the same time they seek to provide an alternative source of economic income for local farmers (Morley *et al.*, 2008). Accordingly, they can be said to distribute more than food. They distribute social connections, relationships and education; in so doing they help to build capacity and increase resilience between local populations of producers and consumers (Franklin *et al.*, 2011). Or at least, these are the principles that guide the development of food hubs. Stroudco is one such initiative that was specifically set up in an attempt to bridge social divisions between food consumers and local producers of food, while at the same time increasing the cross-section of people consuming local food (see www.stroudco.org.uk).

When Stroudco was originally launched in 2009 it was with the aim of providing consumers with an alternative source of local food, the principal products of which were to be produced within 15 miles of the town. The viability of restricting producer membership to this narrow geographical area was supported by data from basic market research, conducted by the founding members in 2008, which had identified that there were some eighty local producers who could potentially be recruited. During the next eighteen months, however, core members of Stroudco experienced a number of setbacks in securing their involvement. The original 'Stroudco Producer

Joining Form' was subsequently realised by the core group to be acting as a significant barrier. In order to trade through Stroudco, producers were initially required to agree, including by written signature, to act in accordance with a set of eighteen Stroudco aims; to abide by a series of forty Stroudco trading and production conditions; and to answer thirty-eight individual questions about their on-farm production and animal husbandry practices, answers to the latter of which it was noted 'may be posted on Stroudco's website and in other publicly available materials'. Included among these were the requirements that producer members would:

- deliver all orders to the trading location (a primary school in the town) on the morning of the trading day – accepting also that because no 'minimum order' applies, even if only a single item was ordered it must still be delivered;
- maintain up to date stock lists on the Stroudco online trading portal;
- offer an event or farm open day relevant to the food production business once a year for consumer members (e.g. picnic, demonstration session, camping, workday, guided farm walk, fruit picking, hay making etc.);
- be willing to stand for election to the Stroudco board every third year;
- undertake a review meeting with the Stroudco manager at least every two years to discuss mutual issues and ideas for improvement.

In response to subsequent recognition by the core group that the producer joining form was overly burdensome and unlikely to be attractive to local farmers, since 2011 the producer joining process has been dramatically simplified. New members now only have to provide basic contact details via an online submission form and then respond to a much reduced number of questions in follow-up as part of a short, informal, telephone conversation with a core Stroudco member. In parallel to this, the requirement that producer members offer an annual event or farm open day, and that they stand for election to the Stroudco board, has been revised from compulsory to voluntary.

Throughout its existence Stroudco has been challenged by what most core group members commonly refer to as a 'chicken and egg situation'. That is, 'without sufficient choice people will not order, but without sufficient numbers of consumers producers will not sign up' (Stroudco core member interview, February 2010). In 2011, in order to remain viable but also retain the interest of current participating producers, the decision was taken by core group members to broaden the Stroudco product range by stocking ethically traded international food stuffs alongside local produce. One of the difficulties in overcoming the challenge of finding other ways to attract a higher number of consumer members was the fact that Stroudco is not the only local food initiative currently operating in Stroud, Gloucestershire. In addition to three other Community Supported Agriculture initiatives, Stroud is also the home of an award winning farmers' market. The success of the farmers' market is reflected in the fact that it has a waiting list of new producer members.

As well as the Stroud farmers' market giving local producers access to a significantly larger consumer base than Stroudco is currently able to offer, it is also potentially more attractive in a number of other respects. In addition to not needing to maintain an online produce stock list, or sell produce to consumers at a rate slightly below normal retail trading price (in line with the Stroudco aim of broadening the social demographic of community members consuming local food), participation in the farmers' market also leaves producers free from any requirement to host an annual event or farm open day, or any perceived obligation to volunteer their time and skills in any other capacity. In effect, therefore, while the farmers' market enables them to build connections with the local consumer base, they are able to do so in a way that, as well as making good economic sense, also retains a separation between consumers and their spaces of food production.

Farmer entrepreneurialism and rural community landshare

Although community food initiatives continue to attract widespread support from a range of stakeholders, deciding whether to permit shared use of an area of land by a community group remains the prerogative of the landowner. As the above example of Stroudco illustrates, however, adopting a much more socially engaged and community-orientated approach to multifunctional agriculture potentially places considerable pressures on the time and resources of farmers. Farmer behaviour in this regard is affected by the degree to which it may be possible to integrate farming activity with rural or, where applicable, peri-urban and urban economies. Integration can be encouraged by those aspects of economic activity that may add value to, and/or act in synergy with the farm and the farmer's resources.

Entrepreneurial activity to support such integration will be affected by opportunities that are accessible to those farmers who are capable and motivated to take advantage of them. Farmer entrepreneurship can, however, often be constrained and narrowly expressed. As discussed above, productivist, value adding or other diversification models lead to differentiated business strategies, which bias the degree to which farmers may be able to participate in collaborative activity with community groups. Farmers with a productivist approach tend to develop, or need, less extensive business networks as they work within relatively rigid hierarchical supply chains, while those that add value to farm products, or diversify their activity necessarily look to more diverse networks. A wider network of contacts from a broader set of market relationships may be expected to improve the farmers' ability to recognise and realise opportunities, particularly those available through non-traditional market actors such as community food hubs, CSAs, recreational and other social interests. Further consideration needs to be given to the significance of farmer entrepreneurship in connection with the range of factors likely to influence a rural landowner's willingness (or otherwise) to make land available for community landshare initiatives. In exploring the potential for increasing the spread of community food growing initiatives beyond the city edge in this way, we do so here by briefly considering the possibility for including more commercially oriented forms of arrangement. We illustrate this by drawing on the non-food example of equestrian livery yards – a rural business opportunity that is firmly based upon multifunctional and multi-user community landshare practice.

Equestrian livery yards provide a good illustration of the fact that, where effectively managed, commercial initiatives based on differing forms of collective landshare practice can be operated in parallel to the more traditional food producing practices of farming. The establishment of livery yards as an alternative or parallel use of farm resources has spread rapidly in urban-fringe areas of the UK in the last fifteen years or so. Livery yards and related equine-based business pursuits (e.g. competition venues, schooling grounds, equine therapy centres) offer farmers and other rural landowners natural avenues towards diversification and participation in the consumption economy. In so doing, collective recreational practices such as these make a complex contribution to the restructuring of rural space and reconnection of rural and urban communities (Franklin and Evans, 2008).

Livery yards have the potential to provide a series of mutual benefits to farmers and horse-owning clients. For the farmer this includes having a regular source of income equivalent to, or in excess of, the average returns secured from other more traditional types of family livestock farming; the opportunity to continue practising land and animal husbandry as well as actively using the physical infrastructure of the farm (unlike, for example, if they switched only to tourism-based holiday accommodation); the potential to run an equestrian livery business in parallel to other forms of agricultural practice; the chance to generate additional income through sale of farm produce such as hay and straw (as feed and bedding for the horses); and opportunities to extend their social connections. For the horse owner, benefits include avoiding the expense of land ownership; becoming part of a community of shared interest; sharing and learning new skills and knowledge in horse care and resource management; having access to large areas of farmland to ride over that otherwise would be out of bounds; and being able to take advantage of existing connections and business arrangements of the farm including veterinary care services and feed merchants.

The multiple opportunities for mutual benefit that come with equestrian livery yard arrangements make this form of community landshare participation attractive to a growing number of farmers. This raises the question of whether farm-based community food growing initiatives hold equal potential for delivering rewards to farmers as well as community groups. Yet, if this is to be achieved in a way that does not ignore the needs of low-income groups – an already existing criticism of many urban community food initiatives (see, for example, Hinrichs, 2000) – then the livery yard model of clients paying a monthly charge to the farmer for services rendered will not necessarily be appropriate. Significant in this context, however, is that few direct financial subsidies are offered to private landowners as an incentive for developing community landshare initiatives. Moreover, being skilled in managed social relations between different group fractions, educating individuals as to the sustainable use of natural resources, or conveying technical information about farming practice and natural resource management in an accessible manner, is not something that farmers are trained for through current systems. This context, together with the wider research review of this chapter, suggests that much more research is required on the benefits for landowners of making land available for community food initiatives. In the concluding section suggestions are made for how such a research agenda could be taken forwards.

Conclusions: exploring the new land-based interfaces

Rural societies and economies are in a process of transition from a post-World War II emphasis on mono-productive activities, to a mix of activities in coexistence. Many of these new rural economic activities are based upon the direct consumption of environmental goods and services that engender new urban–rural interfaces. As part of this transition we have seen new and traditional sectors of rural activity flourish or decline in response. What this socio-economic change represents is an overall societal revaluing of many of the key components of rurality. However, this revaluing is also dependent upon positive social relationships being maintained, expanded and sustained between individual owners of rural space and the growing combination of interests wishing to engage in more meaningful and shared land-based practices. Community food growing initiatives have served as a primary illustration in this chapter of the role of landowners and the relevance of land ownership, to the future trajectory of these new largely urban–rural based relations.

The range of community food initiatives that exist and their ability to operate more or less formally, and often within the context of minimal property rights being afforded to the community group, makes them potentially compatible with multiple types of land ownership and land management arrangements. We have argued throughout this chapter, however, that insufficient attention has thus far been given within current academic and policy debates to the social and economic relationships that flow through these different land ownership–land use combinations. Accordingly, we conclude this chapter with recommendations for two future avenues of research around the combined themes of community landshare, sustainable food practice, and future land availability.

Particularly relevant in the context of progressing the sustainability paradigm pursued in this volume is the role of social relationships between landowners and users in enabling, constraining or, indeed, preventing the spread of alternative food initiatives. Further research is required to help us better understand the socio-spatial geographies of local food initiatives within, as well as between, urban and rural spaces. It is the social relationships between landowners and the varied sustainable and alternative communities-of-interest that also influence the subsequent management of shared use initiatives and the everyday practices that they support. Where existing land tenure and land management arrangements impose severe constraints on the (profitable) establishment of community-of-interest shared land use initiatives, community groups may take the step of becoming freehold landowners themselves. The advantages of doing so, particularly in terms of freedom of use and the protection of community property rights, are well documented (Brown, 2007). Less researched, however, are the wider and longer-term impacts of such buy-outs. As Brown (2007, p. 509) notes, collective forms of property ownership are increasingly being asserted 'on the implicit or explicit assumption that they are better in some way (e.g. more just) than existing arrangements'. But, what is their cultural impact on surrounding communities and existing community identities invested in land (Massey, 2004)? And, to what extent can these neo-productivist shared land use models also be conceptualised as forming a new wave of 'gated communities'? Additionally, what are the implications of the 'messier underbelly' (Brown, 2007) of collective ownership, group decision making, or shared use, for the sustainable management of the natural resource base? Significant here are the skills, knowledge, attitudes and values of participants of local food initiatives, as well as the contested nature of 'community' itself.

A second avenue for future research enquiry begins with the twin questions of: how do existing systems of land tenure, land use regulation and property rights contribute to, or inhibit, the establishment of community-based food initiatives in either a rural or urban setting? And, how are community food initiatives being used to increase the local, more sustainable, productivity of land – while at the same time potentially redefining the meaning of multifunctional land use? Social relationships between landowners and groups of land users remain under-researched in the context of alternative food initiatives, which in turn also limits our understanding of the opportunities that exist to support any scaling up of such initiatives. For example, could a programme of scaling up of sustainable community-based initiatives be maintained within existing patterns and models of land ownership and land regulation, or would it require (or bring about) a substantive transition in the current system of property rights? These are indeed important questions for both current debates about EU CAP reform and national land use planning policies, given, as we argued in the first part of the chapter, that these are only experiencing at best partial reform from their traditional productivist and protectionist priorities. Clearly, the analysis above of community land share initiatives indicates that they can be potentially constrained by these priorities.

What is increasingly clear, though, given the finite nature of land, and also the wider range and demand for ecosystem goods and services that it supports, is that future decisions over the use and rights to this resource are likely to become more contested (Home, 2009). As noted by Home (2009), landowners and government bodies will face mounting pressure to maximise the multifunctional productivity of land. At the same time, they will also likely come under growing pressure to facilitate the participation of multiple users (and even, multiple user groups) in such activities; decisions on land use will increasingly have to take into account both the needs of local community groups and those of civil society more generally. Overall, the rise in community-based initiatives associated with, but not necessarily limited to, sustainable food and energy in both the rural and urban domains will mean that scholars and policy makers will have to radically rethink their spatial assumptions about functional land use and property rights. This radical thinking will need to go far beyond the relatively slow pace of regulatory and policy reform, whether associated with agriculture or spatial planning (e.g. green belt policies). This suggests that the rise of the sustainability paradigm infers a much more innovative, flexible and multifunctional approach regarding questions of what rural and urban greenspace is for and how best to support the more meaningful use of this resource.

Note

1 'Local food initiatives' is used in this chapter as a collective term to refer to a range of local level and/or community-led sustainable food ventures (including, for example, farmers' markets, community gardens, community supported agriculture and community food hubs).

References

Brown, K. (2007) 'Understanding the materialities and moralities of property: Reworking collective claims to land', *Transactions of the Institute of British Geographers*, vol. 32, no. 4, pp. 507–522.

Cohen-Gewerc, E. and Stebbins, R. (2013) *Serious Leisure and Individuality*, Canada: McGill-Queen's University Press.

Daugbjerg, C. and Swinback, A. (2011) 'Explaining the "Health Check" of the Common Agricultural Policy: Budgetary politics, globalisation and paradigm change revisited', *Policy Studies*, vol. 32, no. 2, pp. 127–141.

Daugbjerg, C. and Swinbank, A. (2012) 'An introduction to the "new" politics of agriculture and food', *Policy and Society*, vol. 31, pp. 259–270.

Diakosavvas, D. (ed.) (2006) *The Development Dimension: Coherence of agricultural and rural development policies*, Paris: OECD.

Dwyer, J., Ward, N., Lowe, P. and Baldock, D. (2007) 'European rural development under the common agricultural policy's "second pillar": Institutional conservatism and innovation', *Regional Studies*, vol. 41, no. 7, pp. 873–887.

European Commission (2003) Council Regulation (EC) No 1782/2003; *Official Journal of the European Union*, L 270.

European Commission (2004) 'Fact sheet: New perspectives for EU rural development', Brussels: European Commission.

European Commission (2010) 'The CAP towards 2020: Meeting the food, natural resources and territorial challenges of the future', Communication from the Commission to the European Parliament, the Council, the European Economic and Social Committee and the Committee of the regions, Brussels: COM.

Falk, N. (2006) 'Towards sustainable suburbs', *Built Environment*, vol. 32, no. 3, pp. 225–234.

Federation of City Farms and Community Gardens (FCFCG) (2012) 'Community land for community benefit: Business plan for the Wales Community Land Advisory Service', 5 November. Unpublished report.

Franklin, A. and Evans, R. (2008) 'The creation of new rural economies through place based consumption: Exploring the geographies of equestrian pursuits within the British countryside', BRASS Working Paper Series No. 42, Cardiff: BRASS Centre, Cardiff University.

Franklin, A., Newton, J. and McEntee, J. (2011) 'Moving beyond the alternative: Sustainable communities, rural resilience and the mainstreaming of rural food', *Local Environment*, vol. 16, no. 8, pp. 771–788.

Gill, S.E., Handley, J.F., Ennos, A.R. and Pauleit, S. (2007) 'Adapting cities for climate change: The role of the green infrastructure', *Built Environment*, vol. 33, no. 1, pp. 116–133.

Grant, W. (2012) 'Can political science contribute to agricultural policy?', *Policy and Society*, vol. 31, pp. 271–279.

Hinrichs, C.C. (2000) 'Embeddedness and local food systems: Notes on two types of direct agricultural market', *Journal of Rural Studies*, vol. 16, pp. 295–303.

Hinrichs, C. (2003) 'The practice and politics of food system localization', *Journal of Rural Studies*, vol. 19, no. 1, pp. 33–45.

Home, R. (2009) 'Land ownership in the United Kingdom: Trends, preferences and future challenges', *Land Use Policy*, vol. 26, pp. 103–108.

Marsden, T. (2003) *The Condition of Rural Sustainability*, Assen, Netherlands: Van Grocum.

Marsden, T. and Sonnino, R. (2008) 'Rural development and the regional state: Denying multifunctional agriculture in the UK', *Journal of Rural Studies*, vol. 24, no. 4, pp. 422–431.

Marsden, T., Murdoch, J., Lowe, P., Munton, R. and Flynn, A. (1993) *Constructing the Countryside*, London: UCL Press Ltd.

Massey, D. (2004) 'Geographies of responsibility', *Geografiska Annaler: Series B, Human Geography*, vol. 86, no. 1, pp. 5–18.

Mehmood, A. (2012) 'Urban sustainability: Between mimetics and metaphors', Proceedings of the Sustainability through Biomimicry Conference, pp. 31–48.

Morgan, S.L., Marsden, T., Miele, M. and Morley, A. (2010) 'Agricultural multifunctionality and farmers' entrepreneurial skills: Explorations of the Tuscan and Welsh models', *Journal of Rural Studies*, vol. 26, no. 2, pp. 116–129.

Morley, A., Morgan, S. and Morgan, K. (2008) *Food Hubs: The 'missing middle' of the local food infrastructure?*, Cardiff: BRASS Centre, Cardiff University.

Moyer, H.W. and Josling, T. (2002) *Agricultural Policy Reform: Politics and process in the EU and US in the 1990s*, Aldershot: Ashgate.

OECD (2001) *Multifunctionality: Towards an analytical framework*, Paris: OECD.

Priemus, H. and Hall, P. (2004) 'Multifunctional urban planning of mega-city-regions', *Built Environment*, vol. 30, no. 4, pp. 338–349.

Priemus, H., Rodenburg, C. and Nijkamp, P. (2004) 'Multifunctional urban land use: A new phenomenon? A new planning challenge?', *Built Environment*, vol. 30, no. 4, pp. 269–273.

Ravenscroft, N. and Taylor, B. (2009) 'Public engagement in new productivism', in M. Winter and M. Lobley (eds) *What Is Land For? The food, fuel and climate change debate*, London: Earthscan, pp. 213–232.

Sonnino, R. and Marsden, T. (2006) 'Alternative food networks in the south west of England: Towards a new agrarian eco-economy?', in T. Marsden and J. Murdoch (eds) *Between the Local and the Global: Confronting complexity in the contemporary food sector*. Amsterdam: Elsevier, pp. 299–322.

Sonnino, R., Kanemasu, Y. and Marsden, T. (2008) 'Sustainability and rural development', in J.D. Van der Ploeg and T. Marsden (eds) *Unfolding Webs: The dynamics of regional rural development*, Assen, Netherlands: Royal van Gorcum, pp. 29–52.

Stebbins, R. (1982) 'Serious leisure: A conceptual statement', *The Pacific Sociological Review*, vol. 25, no. 2, pp. 251–272.

Sustain (2013) *Growing Success: The impact of capital growth on community food growing in London*, London: Sustain. Report available at www.sustainweb.org.

Swanwick, C., Dunnett, N. and Woolley, H. (2003) 'Nature, role and value of green space in towns and cities: An overview', *Built Environment*, vol. 29, no. 2, pp. 94–106.

UNHabitat (2009) *Planning Sustainable Cities*, Global Report on Human Settlements 2009, Nairobi: UN Habitat.

Wilson, G.A. (2007) *Multifunctional Agriculture: A transition theory perspective*, Wallingford, Oxford: CABI.

Wilson, G.A. (2008) 'From "weak" to "strong" multifunctionality: Conceptualising farm-level multifunctional transitional pathways', *Journal of Rural Studies*, vol. 25, pp. 367–383.

9 The 'new frontier'?

Urban strategies for food security and sustainability

Roberta Sonnino and Jessica Jane Spayde

Introduction

With more than half of the world's population urbanized, cities find themselves in the forefront of the food sustainability challenge (involving both food security and environmental sustainability). Under the 'New Food Equation' – shaped by food price hikes, dwindling natural resources, social unrest and looming climate change (Morgan and Sonnino, 2010), it is becoming increasingly clear that food insecurity is not simply a problem of insufficient production. Rather, it relates to a complex interaction of structural factors that encompass the entire ecology of the food system (Lang, 2010) and that raise important questions about spatial, economic and cultural access to food (Sonnino, 2009a). Addressing these questions has become an especially urgent priority in cities, where consumers are largely separate from the productive landscape and depend on the market for food (Yngve *et al.*, 2009).

In Europe and North America, pioneering urban governments are devising a new policy and planning approach that aims to address the new food security challenges in a more structural and systemic fashion. As exemplified by the emergence of urban food policy councils and the recent proliferation of urban food strategies, central to this approach is an attention to the complex and interrelated dimensions of the food system that effectively build (or fail to build) opportunities for food security (Ashe and Sonnino, 2013).

The worlds of theory, policy and practice have begun to recognize that there is significant scope for change associated with these urban innovations. As Morgan and Sonnino (2010, p. 210) acknowledge, cities are acquiring a new role – 'namely, to drive the ecological survival of the human species by showing that large concentrations of people can find more sustainable ways of co-evolving with nature and that urban governance is key to creating sustainability changes'. By forging new alliances between food consumers and producers, it has been argued, systemic urban food strategies are creating new forms of connectivity across urban and rural landscapes (Marsden and Sonnino, 2012) that challenge conventional development theories and planning models. Indeed, a recent publication by the FAO on food, agriculture and cities has acknowledged:

> [A] new paradigm is emerging for eco-system based, territorial food system planning [that] seeks ... not to replace the global food supply chains that contribute to food security for many countries, but to improve the local management of food systems that are both local and global.
>
> (FAO, 2011, p. 6)

There is a need to enhance theoretical and political understanding of this new urban foodscape, given its food security and wider sustainability potential.

In this chapter, we focus on the critical role of cities as important centres of change in the food system, and for contributing to a sustainable food paradigm. Based on the comparative analysis of an emerging (but still very fragmented) literature on Food Policy Councils (FPCs) and urban food strategies (UFSs), and primary data collection during observation of several London Food Board meetings, we identify the key innovations that pioneering urban governments are introducing in the food policy arena, and then discuss their effectiveness and real and potential benefits in wider sustainability terms.

The emergence of cities as food policy actors

Agri-food policy has traditionally been fashioned by a more or less formal partnership between national governments and international bodies (such as the WTO), and a narrow and self-referential agri-business sector (Morgan *et al.*, 2006). In the context of food security, this has translated into a global policy approach that has placed too much emphasis on food production, at the neglect of the other fundamental dimension of food security: the accessibility of healthy and nutritious food (Sonnino, 2009a).

The negative consequences of this narrow approach are especially evident in urban contexts, where residents (especially women and those on low income) lack access to the physical assets (such as land and housing) that are necessary to produce food (Bohle and Adhikari, 2002; UNFPA, 2007). Compared to rural populations, who live close to the centres of food production and are often, at least to some extent, involved in it, urban residents must purchase their food, and this is particularly a problem in developing countries, where the poorest citizens spend most of their income on food – as much as 85 per cent in Dar es Salaam, to give one example (Sonnino, 2009a; Redwood, 2009). Simply put, as Bohle and Adhikari (2002, p. 403) state, 'urban food security depends mostly on whether a household can afford to buy enough food'. At a time of volatile food prices and unstable income opportunities (Garrett, 2000, p. 1), 'the city is a critical development frontier and has particular dynamics and cross-scale linkages that need to be considered in order to understand – and ultimately address – the growing epidemic of urban food insecurity' (Crush and Frayne, 2011, p. 540). While food insecurity is occurring in the cities of both developed and developing nations, the scale and scope of the epidemic varies depending on the context.

Throughout the world, pioneering city governments are beginning to see themselves as food system players on the national and international scene, often invoking their public health mandate to address the noxious legacy of lax planning policies that squeezed out urban and peri-urban food producers and enabled the fast food industry to colonize the urban environment (Morgan, 2009). Recent policy documents demonstrate a clear awareness of the unique role that cities are aiming to assume in this realm. In 2010 alone, New York City, for example, aspired to become 'a leader in food systems change' (New York City Council, 2010, p. 3); Los Angeles explicitly aimed to become 'a world leader' in the provision of 'healthy, affordable, fair and sustainable food' (Los Angeles Food Policy Task Force, 2010, p. 6); Toronto stated its reputation 'as a world leader in food thinking and action' (Toronto Public Health Department, 2010, p. 6); and Malmö committed itself 'to lead by example' (Malmö City Council, 2010, p. 23).

From a governance perspective, there are two main mechanisms used by cities to implement change in the food systems: the establishment of Food Policy Councils (FPCs) and the drafting of Urban Food Strategies (UFSs) – documents, such as food charters and plans, that contain a vision statement and an action plan or strategy for a more sustainable food system. There is considerable transformative potential associated with these two mechanisms. Indeed, while on the one hand FPCs have already proven to be effective tools in creating 'comprehensive food system policies' that improve public health and the general quality of life (Muller *et al.*, 2009, p. 225), the emerging urban food strategies are also signalling a new and concrete willingness to help regional farmers who are dwindling in the face of the globalization of the food system, contribute to public health and create new and more sustainable connections between urban, peri-urban and rural environments (Sonnino, 2009b; Marsden and Sonnino, 2012).

Research on these governance innovations is still in its infancy. In general, scholars recognize that the emergence of urban food policies represents 'an important rupture with the past' (Renting and Wiskerke, 2010, p. 1902), but research in this area is still very underdeveloped when compared to housing and health, for example (Smith, 1998; Bohle and Adhikari, 2002). Studies on UFSs have so far focused mostly on urban agriculture (Sonnino, 2009b) and on developing countries (Guyer, 1987; Pryer and Crook, 1988; Bohle and Adhikari, 2002). FPCs, on their part, have attracted some scholarly attention, but, as Schiff (2008, p. 207) notes, 'there still exists a paucity of research on their function (organizational role) in relation to the development of sustainable urban food systems'.

Clearly, there is an increasingly recognized need for what Crush and Frayne (2011, p. 540) define as 'a comprehensive programme of research to create a new urban food security agenda and formulate city-specific security plans'. Central to this agenda is the provision of in-depth studies and comparative analyses that can help policy makers and planners to understand the nature and functioning of the emerging urban and regional foodscape

(Sonnino, 2009b; Wiskerke, 2009). To contribute to this goal, in the remainder of the chapter we will attempt to distil the transformative potential of emerging urban food governance innovations, especially in relation to their organization and structure, the vision and actions proposed, their underlying approach and principles and the instruments available to cities to effect change. In the conclusions, we will propose a guiding framework for sustainable urban food strategies that builds on the various innovations so far introduced by pioneering city-governments in the food policy arena.

Reinventing food governance: food policy councils and urban food strategies

Following the lead of Toronto, the first city to establish its own food policy council (FPC) more than a decade ago (Blay-Palmer, 2009), many North American cities have, over time, adopted this kind of governance arrangement – an idea that has recently spread also to Europe, where Bristol (UK) has been the first city to formalize a FPC. At the national level, policy councils on health and food are not a novelty per se. Indeed, as Lang *et al.* (2005, p. 12) explain, 'they have a long pedigree in Scandinavia', particularly in Norway and Finland, where the first recorded Nutrition Councils were created, respectively, in 1937 and in 1954. In the UK, the House of Commons' (Parliamentary) Health Committee's report on obesity, produced in 2004, concluded that 'national food and health policy lacked coherence, integration and effectiveness', and UK consumer bodies began to lobby for the establishment of a food policy council at the national level (Lang *et al.*, 2005, p. 11).

Broadly speaking, a FPC is an organization of people who are endowed with a mandate and, at least ideally, the power and authority to effect food system change through the design of policies that integrate food with other policy areas – including health, the environment, transport and anti-poverty (Muller *et al.*, 2009, p. 238). In this sense, a FPC aims to provide policy level coherence and communication (or a 'neutral ground') between different governmental functions to aid policy formation – not implementation, which remains within the political administrative framework. As Lang *et al.* (2005, p. 14) state, 'the role of FPCs should not be to duplicate effort, but rather to draw upon, and potentially refocus, efforts already being made'.

Although usually 'the creation of policy councils is in the hands of governments' (Lang *et al.*, 2005, p. 17), in her interviews with FPC members across the USA Schiff found:

> [F]ormal associations with government may restrict the ability of these organizations to propose changes to government structures and policy. In other words, it may be difficult to 'operate within a system and at the same time propose alternatives to that system'.
>
> (Schiff, 2008, p. 216)

In this respect, one of the most innovative aspects of urban FPCs is their heavy reliance on the involvement of civil society. In the case of Bristol, for example, the local civil society[1] was largely responsible for initiating the urban food reform process that eventually culminated with the formation of the FPC. Likewise, in Utrecht (Netherlands), the FPC, which was organized by two environmental NGOs, came to involve a breadth of actors from civil society – including food producers, shopkeepers, cooks, retailers, consumer organizations, consultants and other NGOs (Wiskerke, 2009). In this sense, urban FPCs can be seen as a bottom-up response to national and global policies that, as mentioned above, have failed to provide food system change. Where community groups and NGOs were once content to advocate for 'alternative food systems' from the margins of the political arena, they are now actively collaborating with the local state (Barmeier and Morin, 2012) to address the structural causes of urban food insecurity and unsustainability. In their policy documents, cities often explicitly praise the role of civil society in bringing about positive change. New York City Council, for instance, has emphasized the role that individual consumers, alongside of businesses and municipalities, have historically played in triggering positive changes at the national level (New York City Council, 2010, p. 10). The Toronto Public Health Department has also highlighted 'the role of citizens in shaping the emerging food system' in one of its most recent food strategy documents (Toronto Public Health Department, 2010, p. 7).

Although it is still too early to judge the effectiveness of this governance innovation, which, in many cities, has only recently been introduced, it may be worthwhile here to briefly summarize the different functions that a FPC can perform and that, according to Lang *et al.* (2005, p. 14), no other body could offer. An emerging literature identifies four basic functions for these bodies: to give advice and provide research; to encourage changes in the food system; to engage with stakeholders; and to educate the public.

Giving advice includes many different activities, such as performing scenario building for potential policies and identifying problems (Lang *et al.*, 2005), appraising solutions for the best fit in the context of the city, reviewing the policies, and providing research and background information for decision making (Lang *et al.*, 2005). Giving advice also includes monitoring progress and implementation changes (Lang *et al.*, 2005). For example, the London Food Board (LFB) has given advice by being involved with the London 2012 Olympic and Paralympic Games Food Vision. They were asked to sit on the collaborative group of food stakeholders across the UK to jointly develop a Food Vision for catering at the Games. The LFB, which was one of the key stakeholders in this process (J. Spayde, interview notes, 2012), is also developing a toolkit to assist London boroughs in creating healthier food options around schools (J. Spayde, field notes, 2012).

Encouraging change in the food system includes mobilizing relevant organizations (Lang *et al.*, 2005) and providing policies, standards and funding opportunities (WHO, 1986; Blay-Palmer, 2009) that eliminate reliance on charity-based funding (Caraher and Coveney, 2004; see also Morgan and Sonnino, 2010). The LFB fulfils this function by providing funding for initiatives and by commissioning research. They write supporting letters for NGOs such as Greenwich Co-operative Development Agency (GCDA) and Sustain (J. Spayde, field notes, 2012) and they are involved with a funding initiative, led by the Mayor's office, for the regeneration of high streets. The LFB commissions research by either funding it directly or by finding university students and faculty that are already working on these topics (J. Spayde, field notes, 13 September 2012).

Engaging with stakeholders includes promoting networking between different stakeholders, policy makers and organizations (Lang *et al.*, 2005; Morgan and Sonnino, 2010); involving large business on terms decided by the FPC (Caraher and Coveney, 2004); advocating for stakeholders and for the changes that the FPC is striving to create (WHO, 1986; Lang *et al.*, 2005; Blay-Palmer, 2009; Morgan and Sonnino, 2010); and building consensus among the stakeholders involved (Lang *et al.*, 2005), providing mediation when necessary (WHO, 1986; Blay-Palmer, 2009). Several members of the LFB have emphasized the special role the LFB can play in facilitating exchange and creating networks among London food system actors. During a meeting, one member suggested: 'Let's use the LFB as a place to bring people together to learn about the issues and define the problems we should address.' Another member explained how the LFB is specially situated to bring people together by saying: 'People respond to a letter from the mayor and will come to City Hall [for a symposium or a workshop] if we invite them.' For regular quarterly meetings, the LFB also invite people from NGOs, government departments, companies, and universities to discuss issues, expand their networks, provide assistance, and learn from these stakeholders.

The final important role of FPCs is an *educational* role that involves educating stakeholders on the issues and possible solutions (Morgan and Sonnino, 2010); providing policy learning (Lang *et al.*, 2005); and promoting youth education (Wiskerke, 2009) – in a few words, creating the cultural context necessary for ensuring lasting policy changes. The LFB educates policy makers by providing a forum where Council Members from different boroughs can learn about procurement practices. They also provide city planners with a 'Toolkit on Take-Aways' for how to establish exclusion zones around schools for unhealthy fast food outlets. The LFB also supports the educational efforts of NGOs such as the 'Good Food for London' report, which maps how each borough is progressing on different sustainability initiatives (Sustain, 2011, 2013) (J. Spayde, field notes, 11 October 2012) and supports education efforts by disseminating research results. The LFB members are currently considering writing letters to boroughs to disseminate the research from the caterer Aramark, which demonstrates that boroughs can save money by implementing new or retaining existing public catering programmes, specifically 'Meals on Wheels'. University research is also disseminated by the LFB, as they are writing into a funding bid for regeneration of high streets the results from a study that shows the only sectors experiencing growth during the recession are fast food takeaways, betting shops and pawn stores.

An essential part of an effective FPC is the design of an urban food strategy (UFS) – a document that, at least ideally, contains a vision statement, an action plan and a set of indicators that will allow the city to monitor changes and progress in the transition towards a sustainable food system. Like regional food strategies, these urban documents aim to create synergies and coherence among a variety of activities and roles within the city and between the city and its surrounding rural hinterland (see Brunori and Rossi, 2000). An analysis of the strategies devised so far reveals that the overall vision that frames the urban discourse on food invariably emphasizes its multifunctionality – that is, its multiple and synergistic relationships with human and environmental health. Brighton, for instance, one of the first UK cities to take action in the food policy realm, makes specific reference in its overarching vision to the relationship that the food system has with 'social equity, economic prosperity, environmental sustainability, global fair-trade and the health and well-being of *all* residents' (Brighton and Hove Food Partnership, 2012, italics in original). Newquay, also in the UK, refers to the connections between food, health, the environment and economic regeneration in its definition of the main goal of its food strategy. For the Delaware Valley Regional Planning Commission (DVRPC, 2011) in Philadelphia, food holds the potential to strengthen the agricultural sector, decrease waste, improve public health, protect soil and water resources and encourage 'diversity, innovation and collaboration'.

Against this shared background, cities however propose a different central narrative in their food strategies: some cities target primarily their urban food economy; others prioritize the needs, especially health needs, of their residents. New York is an example of an economically driven food strategy. On the basis of its significant purchasing power, the city sees itself as 'uniquely positioned to stimulate the food economy, strengthen our regional food system, and drive local and regional business activity' (New York City Council, 2010, p. 3). Similarly, Bristol's food strategy identifies as a primary objective the creation of a 'sustainable and resilient food economy' that can support the health of its communities and of the environment. Vancouver and Leicester both identify the improvement of the financial viability of their food sector as a top priority, whereas Philadelphia goes as far as defining the local and healthy food movements as 'economic development strategies' (DVRPC, 2011, p. 4).

Among the cities that give priority to citizens' health in their narratives, it is worth mentioning Toronto's visionary idea of a 'health-focused food system' that addresses all factors that influence the health of individuals, families, neighbourhoods and cities. As the Canadian strategy points out, this means more than making safe and nutritious food available to urban residents: a health-focused food system 'nourishes the environment, protects against climate change, promotes social justice, creates local and diverse economic development, builds community' (Toronto Public Health Department, 2010,

p. 6). Los Angeles and Malmö both utilize the notion of 'good food' to emphasize the centrality of citizens' health (also in relation to other sustainability objectives) in their food strategies. In the American city's document, a 'good food' system is defined as one that 'prioritizes the health and well-being of our residents; makes healthy, high-quality food affordable; contributes to a thriving economy ...; protects and strengthens our biodiversity and natural resources throughout the region' (Los Angeles Food Policy Task Force, 2010, p. 11). Other FPCs, such as the LFB, attempt to support both the economic and health aspects of food by attaching food to the regeneration of high streets initiative, focusing on creating a 'Good Jobs in Food' apprenticeship programme, while simultaneously aligning with the mayor's health team initiatives on obesity and food poverty (J. Spayde, field notes, 2012).

Despite these differences in the central food narrative, all UFSs see re-localization as crucial (at least to some extent) to the achievement of their central objective – be it the strengthening of the urban food economy or an improvement in the health of residents. Significantly, 'localness' is in no instance confined to the municipal boundaries. Indeed, although most strategies recognize the importance of increasing urban food production through support for urban agriculture and growing schemes, these initiatives do not have the centrality that academic literature has often accorded to them. Indeed, in an era of increasing food insecurity, cities aim, first and foremost, at re-establishing more sustainable environmental, social and economic relations with their surrounding countryside.

Los Angeles is one of the most explicit advocates of this broad and flexible interpretation of food re-localization. As stated in its strategy:

> [W]hile the benefits of urban agriculture are significant to individuals and neighbourhoods, poverty and hunger ... exist on such a massive scale that supporting urban agriculture should only be viewed as a *supplement, not a replacement*, strategy to solve food insecurity and improve food access.
> (Los Angeles Food Policy Task Force, 2010, p. 26, italics added)

Other cities, such as New York for instance, make a similar point by distinguishing between local/urban and local/regional food systems – or, as stated in Oakland's food strategy, between food 'produced within the city' and food 'coming from the regional local foodshed' (Unger and Wooten, 2006, pp. 12–13).

What exactly is a local food system, as defined and envisioned by cities in their policies and strategies? North American strategies generally make more of an effort to define a 'regional–local', either by equating it with the State in which they are located (as is the case for Vancouver) or, more often, by referring to the notion of 'foodshed' – a geographic area of 100 miles from which, as Philadelphia's strategy states, a population's food 'may theoretically be sourced' (DVRPC, 2011, p. 4). Significantly, cities often attach to this notion a series of specific qualities in terms of agricultural production methods employed, fair farm labour practices, environmental indicators and animal welfare (Los Angeles Food Policy Task Force, 2010, p. 17; see also Thompson *et al.*, 2008). Local, then, is more than a spatial scale. In essence, it is viewed and promoted as a means to an end – that is, as a tool to enhance economic development and employment opportunities for urban residents (Brighton and Hove Food Partnership, 2006, pp. 2–4), increase the availability of fresh and nutritious food (ibid., p. 4) and promote environmental conservation (ibid., p. 11) – through, for example, increased protection of biodiversity (Malmö City Council, 2010, p. 25) or decreased energy costs (New York City Council, 2010, p. 4).

UFSs in general mention three principles that support and promote relocalization: justice, control and environmental conservation. Many UFSs have a strong focus on *justice* and rights, including the right for every citizen to have access to healthy and nutritious food – a principle embedded in the general notion of 'good food' mentioned above. The items specified in UFSs include ensuring food citizenship, social justice and food access (Lang *et al.*, 2005; Friedmann, 2007; Lang and Rayner, 2007; Wiskerke, 2009), as well as addressing the problem of food insecurity (Morgan and Sonnino, 2010). Related to this, many UFSs address the issue of global corporate *control* of the food chain (Welsh and MacRae, 1998) and the local de-skilling and isolation caused by the global food system (Caraher and Coveney, 2004). In this respect, UFSs create mandates for bringing control back to the local area by prioritizing the creation of local market opportunities and jobs (Lang *et al.*, 2005; Friedmann, 2007; Lang and Rayner, 2007; Wiskerke, 2009; Morgan and Sonnino, 2010) and by creating 'buy local' standards for public procurement managers (Morgan and Sonnino, 2010).

With regard to *environmental conservation*, UFSs often prioritize local agriculture, including rural, regional, peri-urban and urban agriculture (Wiskerke, 2009) and, more generally, conserving agricultural land, including for environmental and cultural heritage reasons (Lang *et al.*, 2005; Friedmann, 2007; Lang and Rayner, 2007; Wiskerke, 2009; Morgan and Sonnino, 2010). As an additional environmental problem, some UFSs also address waste issues, including separating food and municipal waste streams for compostable items (Morgan and Sonnino, 2010).

Towards an integrated policy approach: the transformative potential of urban food governance

Academic researchers agree that finding sustainable solutions for the food system necessitates cross-disciplinary and multifunctional approaches that steer away from compartmentalized thinking. As Rayner (2009, p. 590) puts it, 'the problem is not just the dualistic separation of natural and human ecology, but fragmentation and rigidities of all kinds' that overlook 'the complex interdependencies between socio-economic and policy forces', creating a thicket of poorly understood drivers and unclear policy options' (cited in

Muller *et al.*, 2009, p. 231) for the food system. Recognizing these complex interdependencies is crucial to address problems structurally and systematically – or, in other words, to devise a meaningful and effective approach to food policy making (Muller *et al.*, 2009, p. 231).

In theory, embracing the complexity of the food system means integrating ecology and social thinking (Lang, 2005), moving away from the individualistic, linear and mechanistic thinking that emphasizes consumer choices, diverting attention from the real causes and determinants of people's lack of access to healthy food – namely, history, class, gender, income, ethnicity, affordability and global supply patterns (Caraher and Coveney, 2004). Practically, this entails structuring food policies around an explicit recognition of the multidimensional connections that food has with different social contexts (including the family and labour markets) and with other community systems such as housing, transport, land use and economic development (Blay-Palmer, 2009; Sonnino, 2009a).

UFSs are perhaps the first tangible example of this systematic, structural and integrated policy approach to the food system, which, in some cases, is explicitly praised by city governments. Los Angeles, for example, advocates the integration of local food system planning into its Regional Climate Action Plan and Transportation Plan (Los Angeles Food Policy Task Force, 2010, p. 60). Similarly, Newquay's strategy makes clear that healthy and sustainable food considerations have a significant contribution to make towards the objectives of the local Sustainability Strategy – namely, limiting the population's greenhouse gas emissions and ecological footprint and promoting regional economic development (Duchy of Cornwall, 2009, pp. 7–8). The benefits and scaling up potential of this new policy approach are becoming increasingly evident to cities, especially those that pioneered the design and implementation of UFSs. In Toronto, for example, municipal authorities have already designated as a 'pressing strategic opportunity' the possibility of linking 'the evolving food cluster with the developing green economy and market[ing] the region as the continent's go-to-region for food innovation' (Toronto Public Health Department, 2010, p. 8). For New York City, in turn, the scope for scaling up this emerging policy innovation is broader than the urban–regional level. Indeed, the American city has the explicit ambition of becoming a laboratory of food system innovation – or, as stated in its strategy, 'a model of how targeted local action can support large scale improvements' (New York City Council, 2010, p. 3).

In addition to promoting a new integrated vision for food policies, UFSs and FPCs are building new networks in the food system – 'new links and new relationships between different stages and actors of the food chain' (Sonnino, 2009a, p. 431). As Lang *et al.* (2005) argue, FPCs are an especially useful mechanism to address the 'failures of coordination' (Lang *et al.*, 2005, p. 12). Indeed, they act as 'networkers across the spectrum of food system interests and facilitators in the networking and implementation capacity of other organizations' (Schiff, 2008, p. 216). They can also facilitate dialogue (Blay-Palmer, 2009), bring together stakeholders (Kim and Blanck, 2011) and create avenues for 'alliances and lobbying' (Caraher and Coveney, 2004). Finally, FPCs can fulfil a networking role by creating or enhancing community coalitions that involve 'multiple sectors of the community coming together to address community needs and solve community problems' (Berkowitz and Wolff, 2000). For Lewis *et al.* (2011), a coalition is an 'essential vehicle for engaging the community, expanding community leadership, and diffusing knowledge' (p. 95). This is especially the case when a FPC is effectively resourced and includes members who represent a breadth of different interests and expertise – factors that enable the FPC to play a facilitation role. An illustrative example here is Toronto, where the diversity of experience of the board members placed the FPC in the position of providing 'authoritative, credible input on an extensive range of food-related issues' (Blay-Palmer, 2009, p. 405). In the Canadian city, networking was fuelled by a C$2.4 million grant financing over 150 projects, which facilitated the creation of many different kinds of formal and informal networks (some with staff financed by the grant programme) that provided 'tangible evidence of the rooting of a food citizenship and community food security agenda' in the city (Welsh and MacRae, 1998, p. 253). The LFB also helps facilitate long-lasting networks by designing an apprenticeship programme that will provide food workers with a holistic view of the food chain and establish the necessary networks to support decision making that takes into account the interests of different actors in the food chain (J. Spayde, field notes, 2012).

Facilitating these networks is not always easy for FPCs. Schiff found that they often 'need to strike a balance between authority within government, freedom of expression, and ease of communication among a wide variety of food system stakeholders' (Schiff, 2008, p. 216). But there are important benefits that spur from enhanced networking capacity in the food system. In addition to providing support for local businesses, FPCs can also help organizations to gain political capital (Schiff, 2008), which is important especially for alternative, non-profit, disadvantaged or neighbourhood food organizations that have suffered from a lack of such capital. The existence of a FPC can provide an opportunity for organizations to 'gain some of the much-needed or sought after credit for [the] projects' they have been involved in (Schiff, 2008, p. 218). Additionally, many FPCs have the specific goal to enhance what other groups who have similar goals are doing, rather than competing with them (Schiff, 2008, p. 218).

Finally, UFSs and FPCs also provide a platform where city issues and problems can be made more visible. As sociologist C. Wright Mills (1959, p. 8) explains, a problem is not seen as a 'social problem' to be addressed collectively by society if it is not presented as such. For instance, as mentioned earlier, a healthy diet can be illustrated as a personal issue of choice and knowledge, or it can be constructed as a social problem that relates to the underlying cultural and economic structure. In a context where problems such as diet-related diseases, food access and food choices are popularly perceived as individual problems, UFSs and FPCs have an important role to play in terms of 'raising the visibility of a broad spectrum of food system interests in government policy, planning and decision-making activities' (Schiff, 2008, p. 216). Commonly recognized at LFB meetings, for example, is that the Mayor's support, in the form of a letter or a mention in a speech, is a highly valuable asset in legitimizing food system issues in London (J. Spayde, field notes, 2012).

In short, as Blay-Palmer (2009) argues in her analysis of Toronto's FPC, this organization 'provides an example of how urban spaces – in physical and policy terms – can be used for progressive ends to create "the just city"' (Blay-Palmer, 2009, p. 401) – or, as theorized by Fainstein (2006), a city that merges aspects of the built environment with progressive social, environmental and economic spaces (including, for example, heterogeneous public spaces; historical and cutting-edge architecture; space for social resistance and conflict; inclusive, mixed development with affordable housing; an economic development that includes small businesses and cooperatives; and environmental regulation and green development). By providing this platform for visibility, UFSs and FPCs ultimately incorporate into the collective conscious of the city ideas of justice that embrace, at the same time, the good health of urban residents and the plight of small local farmers who are increasingly marginalized or even excluded from an increasingly globalized and price-driven food economy.

Tapping the potential of urban food governance

There are three, somewhat interrelated, strategies that cities advocate to progress their integrated food policy approach: infrastructural development; a more enabling planning system; and public procurement. Infrastructural development is especially an issue in the North American cities, which often raise the need for reconnecting local food producers with urban consumers through the development of alternative retail outlets such as farmers' markets and Community Supported Agriculture schemes (which are prioritized by Toronto and Los Angeles, in particular) or the establishment of regional food hubs and permanent wholesale markets (as proposed by New York and Los Angeles).

Spatial planning also figures prominently among the initiatives proposed to increase the supply of and demand for local food products. As stated in the strategies of New York, Philadelphia, Leicester, Manchester, London and Newquay, supportive land ordinances can play a major role in facilitating the production and distribution of local foods in urban and peri-urban areas. In addition to promoting urban food production, imaginative cities are deploying the power of planning to re-regulate other aspects of the urban foodscape – for example, to protect the accessibility of a range of food retail outlets, discourage food waste and create income for producers who need access to the 'footfall' of urban consumers (White and Natelson, 2012).

Some London boroughs have successfully banned fast food outlets within 400 metres of schools, an initiative that the LFB has strongly supported as demonstrated by its effort to develop a toolkit to teach planners across London how to ban takeaways (J. Spayde, field notes, 2012).

Public procurement is the most widely mentioned tool that cities have at their disposal to calibrate demand and supply of healthy food products. In addition to being praised for its potential to contribute to public health, climate change mitigation and regional development (see, for example, the strategies produced by Malmö, Toronto, New York and Philadelphia), sustainable public procurement is often extolled for its integrative potential – or, as stated in the Bristol food strategy (Bristol Food Network, 2009, p. 2), its capacity to foster collaboration between urban communities and the food producers, processors and suppliers located in urban and peri-urban areas. For the authors of Philadelphia's strategy, the type of collaborative relationships that can be forged through sustainable public procurement initiatives are even broader, embracing, as they do, 'all food system stakeholders, ranging from the private sector to the public sector, from local food advocates to hunger relief organisations, from farmland preservation coordinators to economic development agencies' (DVRPC, 2011, p. 11). Some of the most pioneering cities also attach an important educational dimension to the public provisioning of food. This is the case, for example, of Malmö, where public procurement is used to promote the public health agenda and embed public canteens in the new urban foodscape (Malmö City, 2010). In Rome, the long process of school food reform was also supported through the implementation of educational activities that aimed to create a new generation of knowledgeable consumers (Sonnino, 2009b).

As outlined in Chapter 4 of this publication, public procurement is a powerful instrument to reconnect food producers and consumers, as well as rural and urban areas, in the context of what appears to be a generalized urban effort to transcend conventional dichotomies and spatial, sectoral and political separations. As Toronto's food strategy explicitly states:

> [S]ometimes, both the local food movement and its detractors have become absorbed in debates expressing the same compartmentalized thinking that characterizes the dominant food system. … The issue is not so much which single food choice is 'best', but how can we accelerate progress towards a … food system where goals of affordability, environmental protection, local farm viability, land use planning and others can be reconciled.
>
> (Toronto Public Health Department, 2010, p. 12)

The LFB has made a point of using procurement as a tool for changing the food system by supporting existing catering standards (Health Catering Commitments and Food for Life Partnership) and initiating changes in government procurement (i.e. Transport for London) (Morgan and Sonnino, 2008).

Sustainable urban food strategy: a guiding framework for practitioners

The chapter has so far discussed in broad and theoretical terms the main innovations introduced by the recent development of UFSs and FPCs – namely, their emphasis on a new, multifunctional vision of food, their capacity to foster networking within and between food systems, and their potential to give visibility to (and emphasize the social dimension of) food-related issues and problems. In this final section, and based on a comparative analysis of best practices and of the effectiveness of UFSs and FPCs implemented so far, we distil six main criteria that could provide a guiding framework for city governments that aim to reform their urban foodscape.

The first is the *clarity of the overall vision*. An urban policy initiative needs a clear vision that highlights the ways in which food is interconnected with many different policy areas and social issues – including health, transport, education, zoning, social justice and environmental integrity. A clear vision, informed by 'multi-sectoral, cross-disciplinary thinking', can generate widespread support for the strategy, while at the same time providing a principled backbone to re-evaluate and reformulate the action plan in light of wider changes in the food system.

Second, a real reform of the urban foodscape requires the adoption of a *comprehensive policy approach* that considers food changes from farm to fork. Re-localizing the urban food system per se is not a sustainable strategy if a vast percentage of the food that enters the city is actually wasted at other stages of the supply chain. As Caraher and Coveney (2004) convey, the policy approach has to be, at the same time, 'downstream' and 'upstream' to embrace the many complex (and often unsustainable) linkages between different stages of the food chain.

Third, to be effective, UFSs should explicitly and clearly raise the *need for change* and propose a timeline to achieve it. In general, a strategy is more likely to gain support by beginning with small, incremental and less controversial changes, but, as Muller *et al.* (2009, p. 238) state, today 'the momentum of the industrial food system model is too strong to be effectively altered by incremental policies'. In some contexts, urban food reform may need to work its way up to larger, more controversial, radical food system changes. In any case, the distinction between radical and incremental changes helps in setting up different timelines for the goals.

The fourth guiding principle behind the reform of the urban foodscape has to do with the *involvement of stakeholders*. As discussed above, stakeholders have often been involved in the writing of the policy documents and in the formation of FPCs, but very few cities have created mechanisms for ongoing collaboration and joint policy making (Toronto is a notable exception here). As Morgan (forthcoming) argues, at a time when fiscal and financial austerity is making it increasingly difficult for urban governments to meet the basic needs of their populations alone, collaboration (in the form of a new alliance between municipal authorities, social enterprises, civil society groups and the private sector) is an essential aspect of an effective food strategy. As we have described above, the role of NGOs has thus far been especially crucial in orchestrating the arrangements needed to initiate reform processes and in influencing local state policies that mostly bear on the urban food system. In the UK, for example, Sustain's *Good Planning for Good Food* (Sustain, 2011) has helped urban planners to re-imagine their role as enablers of projects that foster food security.

The fifth principle is establishing *mechanisms to evaluate and monitor progress* in food system reform. Monitoring and evaluation involve recognizing when strategies and implementation are successful or not as well as communicating best practices and success to the public to ensure their involvement and support. Much of what is learned through the application of a rigorous and sophisticated monitoring/evaluation system can and should be used also to promote knowledge exchange between city governments, which have been clearly raising the need to share ideas and best practice – as demonstrated, for example, by the establishment of a Sustainable Food City Network, facilitated by the Soil Association, in the UK, where urban food reform is especially high on the policy agenda.

Finally, in the long run, the success of an urban food reform process depends on its capacity to harness *cultural change* through the implementation of reforms that widen citizens' access to healthy food products and strengthen the image of food and its developmental potential. In the UK, for example, the city of Plymouth has built on the Soil Association's Food for Life initiative to deliver a school food service that teaches children and caterers the merits of a whole school approach to sustainable food provisioning (Soil Association, 2012). This example gives evidence of the ways in which urban-led food policy initiatives can help create changes to broaden and deepen nutrition benefits, as with a whole school approach that links the message of the classroom with that of the canteen. Linking a city strategy to a national programme such as Food for Life can significantly increase the reach of the cultural change that sustainable catering initiatives purport to achieve.

It is still too early to judge the impacts and medium- to long-term effects of the urban food reform processes we have discussed in this chapter. But the stories of the many cities that are attempting to re-imagine their urban foodscapes highlight that, regardless of the tangible outcomes of these local policy initiatives, there is a new realization that the current 'sectorialized' farm and food system is not fit for purpose. A vision is emerging here for a new policy approach that emphasizes the multifunctional values of food and its capacity to create more sustainable spatial, economic and sociocultural linkages between urban and rural areas. In a global economic context in which 'deficits and surpluses create new food equations and disrupt established spatial fixes' (Marsden and Sonnino, 2012, p. 3), this vision is an important call for action for all policy makers and researchers working to progress the sustainability agenda.

Moving forward: a research framework for sustainable UFS and FPC

The academic literature on UFS/FPC is still lacking a critique of these emerging sustainable food system initiatives. The following presents a framework for progressing research in this important area, focusing on the concept of embeddedness. We attempt to raise theoretical and empirical questions that can contribute to a better conceptualization of initiatives such as UFSs and FPCs. Strategies and councils can be socially and politically embedded.

One main theoretical suggestion is examining the level and degree of social embeddedness of UFSs and FPCs, which raises questions about the nature of social capital associated with these initiatives. To what extent do UFSs and FPCs engage with civil society? What segments of civil society are involved with the UFS or FPC, and which ones are excluded? What is the relation between food system activists and the UFS or FPC? Does the UFS or FPC have widespread community support? The main point here is that without the involvement and participation of civil society, political initiatives are bound to be partial and fragile.

Another related theoretical question is about the political and institutional embeddedness of these initiatives within a multilevel governance polity. Situating these initiatives in their wider political and institutional context is crucial to understand the kind of support they need from different levels of government and forms of governance. Equally important here is the issue of policy integration within the urban polity. How do these initiatives become embedded within a city's institutional and governance framework? How does the process of institutionalization affect the capacity of an urban food policy initiative to change the food system? As we have argued in this chapter, integrating urban policies to address food from a multiplicity of different departments and subjects is often a key desired outcome of UFSs and FPCs. Therefore, an important question becomes: What is the scope for integrating these policy/governance mechanisms into the urban system so as to account for and enhance the multifunctional values of food and realize its full development potential?

Beyond the city, there are also important questions about the incorporation of urban food policies into a national or regional framework, which can facilitate knowledge exchange and the dissemination of best practice. One example of a national framework is the UK's emerging Sustainable Food Cities Network, a partnered effort of Cardiff University and NGOs Soil Association, Sustain, and Food Matters. As a part of a large network of this kind, which currently includes more than twenty cities, urban governments enhance their capacity to lobby national governments to change policies and obtain funding needed to support sustainable food systems.

The emergence of these regional and national networks brings up another important issue that deserves scholarly attention: that is, how to scale up UFS and FPC initiatives. An important question for theorists to consider is whether UFSs and FPCs are creating a fundamental shift in the food system (as outlined in Flynn and Bailey, Chapter 5, this volume), through, for example, the creation of a new vision for the future of food and the development of innovative and more inclusive governance arrangements, or whether they are merely examples of niches that fail to make a dent in the dominant food discourse and practices. A careful consideration of their social, political and institutional embeddedness has much to say about their inclusiveness and long-lasting potential. Remaining theoretically critical will help ensure that scholars continually sidestep the tendency to oversimplify and uncritically glorify UFS and FPC to progress a research agenda that captures the relational nature of the emerging urban foodscape and its potential to forge new and more sustainable spatial, cultural and socio-economic linkages between cities and their surrounding regions.

Note

1 Such as the non-profit organizations: Bristol Food Network and the Soil Association.

References

Ashe, L. and Sonnino, R. (2013) 'At the crossroads: New paradigms of food security, public health nutrition and school food', *Public Health Nutrition*, vol. 16, no. 6, pp. 1020–1027.

Barmeier, H. and Morin, X.K. (2012) 'Resilient urban community gardening programmes in the United States and municipal-third sector "adaptive co-governance"', in A. Viljoen and J.S.C.Wiskerke (eds) *Sustainable Food Planning: Evolving theory and practice*, Wageningen: Wageningen Academic Publishers, pp. 159–172.

Berkowitz, W.R. and Wolff, T. (2000) *The Spirit of the Coalition*, Washington, DC: American Public Health Association.

Blay-Palmer, A. (2009) 'The Canadian pioneer: The genesis of urban food policy in Toronto', *International Planning Studies*, vol. 14, pp. 401–416.

Bohle, H.G. and Adhikari, J. (2002) 'The metropolitan food system of Kathmandu – Conceptual considerations and empirical evidence', *Erde*, vol. 133, pp. 401–421.

Brighton and Hove Food Partnership (2006) *Spade to Spoon: Making the connections – A food strategy and action plan for Brighton and Hove 2006*, Brighton and Hove: Brighton and Hove Food Partnership.

Brighton and Hove Food Partnership (2012) *Spade to Spoon: Digging deeper: A food strategy and action plan for Brighton and Hove*, Brighton and Hove: Brighton and Hove Food Partnership.

Bristol Food Network (2009) *A Sustainable Food Strategy for Bristol and Bristol Food Network*. Bristol: Bristol Food Network.

Brunori, G. and Rossi, A. (2000) 'Synergy and coherence: Some insights from wine routes in Tuscany', *Sociologia Ruralis*, vol. 40, no. 4, pp. 409–423.

Caraher, M. and Coveney, J. (2004) 'Public health nutrition and food policy', *Public Health Nutrition*, vol. 7, pp. 591–598.

Crush, J.S. and Frayne, G.B. (2011) 'Urban food insecurity and the new international food security agenda', *Development Southern Africa*, vol. 28, pp. 527–544.

DVRPC (2011) 'Eating here: Greater Philadelphia's food system plan', Delaware, WI: Delaware Valley Regional Planning Commission.

Duchy of Cornwall (2009) 'Newquay growth area food strategy', Newquay Growth Area, Newquay: Duchy of Cornwall.

Fainstein, S. (2006) *Planning and the Just City: Searching for the just city*, Columbia: Graduate School of Architecture, Planning and Preservation, Columbia University.

FAO (2011) *Food, Agriculture and Cities: Challenges of food and nutrition security, agriculture and ecosystem management in an urbanizing world*, Rome: FAO.

Friedmann, H. (2007) 'Scaling up: Bringing public institutions and food service corporations into the project for a local, sustainable food system in Ontario', *Agriculture and Human Values*, vol. 24, pp. 389–398.

Garrett, J.L. (2000) *Achieving Urban Food and Nutrition Security in the Developing World*: Overview, Focus 3, Brief 1, August. Washington, DC: International Food Policy Research Institute.

Guyer, J. (1987) *Feeding African Cities: Studies in regional social history*, Manchester: Manchester University Press.

Kim, S.A. and Blanck, H.M. (2011) 'State legislative efforts to support fruit and vegetable access, affordability, and availability, 2001 to 2009: A systematic examination of policies', *Journal of Hunger and Environmental Nutrition*, vol. 6, pp. 99–113.

Lang, T. (2005) 'Food control or food democracy? Re-engaging nutrition with society and the environment', *Public Health Nutrition*, vol. 8, pp. 730–737.

Lang, T. (2010) 'Crisis? What crisis? The normality of the current food crisis', *Journal of Agrarian Change*, vol. 10, pp. 87–97.

Lang, T. and Rayner, G. (2007) 'Overcoming policy cacophony on obesity: An ecological public health framework for policymakers', *Obesity Reviews*, vol. 8, pp. 165–181.

Lang, T., Rayner, G., Rayner, M., Barling, D. and Millstone, E. (2005) 'Policy councils on food, nutrition and physical activity: The UK as a case study', *Public Health Nutrition*, vol. 8, pp. 11–19.

Lewis, L.B., Galloway-Gilliam, L., Flynn, G., Nomachi, J., Keener, L.C. and Slone, D.C. (2011) 'Transforming the urban food desert from the grassroots up: A model for community change', *Family and Community Health*, vol. 34, pp. 92–101.

Los Angeles Food Policy Task Force (2010) 'The good food for all agenda: Creating a new regional food system for Los Angeles', Los Angeles, CA: The Los Angeles Food Policy Task Force.

Malmö City (2010) *Policy for Sustainable Development and Food*, Malmö, Sweden.

Marsden, T. and Sonnino, R. (2012) 'Human health and wellbeing and the sustainability of urban–regional food systems', *Current Opinion in Environmental Sustainability*, vol. 4, pp. 427–430.

Mills, C.W. (1959) *The Sociological Imagination*, New York: Oxford University Press.

Morgan, K. (2009) 'Feeding the city: The challenge of urban food planning', *International Planning Studies*, vol. 14, pp. 341–348.

Morgan, K. (forthcoming) 'The new urban foodscape: Planning, politics and power', in A. Viljoen and K. Bohn (eds) *Urban Agriculture as Second Nature*, London: Routledge.

Morgan, K. and Sonnino, R. (2008) *The School Food Revolution: Public food and the challenge of sustainable development*, London: Earthscan.

Morgan, K. and Sonnino, R. (2010) 'The urban foodscape: World cities and the new food equation', *Cambridge Journal of Regions Economy and Society*, vol. 3, pp. 209–224.

Morgan, K., Marsden, T. and Murdoch, J. (2006) *Worlds of Food: Place, power, and provenance in the food chain*, Oxford: Oxford University Press.

Muller, M., Tagtow, A., Roberts, S.L. and Erin, M. (2009) 'Aligning food systems policies to advance public health', *Journal of Hunger and Environmental Nutrition*, vol. 4, pp. 225–240.

New York City Council (2010) *FoodWorks: A vision to improve NYC's food system*, New York: New York City Council.

Pryer, J. and Crook, N. (1988) *Cities of Hunger: Urban malnutrition in developing countries*, Oxford: Oxfam Publishing.

Rayner, G. (2009) 'Conventional and ecological public health', *Public Health*, vol. 123, pp. 587–591.

Redwood, M. (ed.) (2009) *Agriculture in Urban Planning: Generating livelihoods and food security*, London: Earthscan.

Renting, H. and Wiskerke, H. (2010) 'New emerging roles for public institutions and civil society in the promotion of sustainable local agro-food systems', Presentation given at the *9th European IFSA Symposium*, Vienna, Austria.

Schiff, R. (2008) 'The role of food policy councils in developing sustainable food systems', *Journal of Hunger and Environmental Nutrition*, vol. 3, pp. 206–228.

Smith, D.W. (1998) 'Urban food systems and the poor in developing countries', *Transactions of the Institute of British Geographers*, vol. 23, pp. 207–219.

Soil Association (2012) *Sustainable Food Cities Network*, Soil Association, available at www.soilassociation.org/sustainablefoodcities, accessed 1 October 2012.

Sonnino, R. (2009a) 'Feeding the city: Towards a new research and planning agenda', *International Planning Studies*, vol. 14, pp. 425–435.

Sonnino, R. (2009b) 'Quality food, public procurement, and sustainable development: The school meal revolution in Rome', *Environment and Planning A*, vol. 41, pp. 425–440.

Sustain (2011) *Good Planning for Good Food: How the planning system in England can support healthy and sustainable food*, London: Sustain.

Sustain (2013) *Good Food for London 2013: London Borough maps of progress on healthy and sustainable food*, London: Sustain.

Thompson, W., Meyer, S.D. and Westhoff, P.C. (2008) 'Model of the US ethanol market', FAPRI-MU Report Series 37971, Food and Agricultural Policy Research Institute, University of Missouri.

Toronto Public Health Department (2010) 'Cultivating food connections: Toward a healthy and sustainable food system for Toronto', Toronto: Toronto Public Health Department.

Unger, S. and Wooten, H. (2006) 'A food systems assessment for Oakland, CA: Toward a sustainable food plan', Oakland, CA: Oakland Mayor's Office of Sustainability and University of California, Berkeley, Department of City and Regional Planning.

UNFPA (2007) *State of the World Population 2007: Unleashing the potential of urban growth*, Geneva: United Nations Population Fund.

Welsh, J. and Macrae, R. (1998) 'Food citizenship and community food security: Lessons from Toronto, Canada', *Canadian Journal of Development Studies*, vol. 19, pp. 237–255.

White, H. and Natelson, S. (2012) 'Good planning for good food: Mechanisms within the English planning system to support sustainable food and farming', in A. Viljoen and J.S.C. Wiskerke (eds) *Sustainable food planning: Evolving theory and practice*, Wageningen: Wageningen Academic Publishers, pp. 507–516.

Wiskerke, J.S.C. (2009) 'On places lost and places regained: Reflections on the alternative food geography and sustainable regional development', *International Planning Studies*, vol. 14, pp. 369–387.

WHO (1986) Ottawa Charter for Health Promotion. *International Conference on Health Promotion, 'The move towards a new public health'*. World Health Organization with Canadian Public Health Association and Health and Welfare Canada, Ottawa ON.

Yngve, A., Margetts, B., Hughes, R. and Tseng, M. (2009) 'Food insecurity – not just about rural communities in Africa and Asia', *Public Health Nutrition*, vol. 12, pp. 1971–1972.

10 Conclusions

Building the food sustainability paradigm: research needs, complexities, opportunities

Terry Marsden

Introduction: food as part of world ecology

We have argued in this volume that it is becoming increasingly necessary for agri-food scholarship, and environmental social science scholarship more generally, to adopt a more normative, critical sustainability paradigm in its work and perspectives. As we outlined in Chapter 1 of this volume, this is not to be conceived of as some sort of rigid Kuhnian (1962) 'straight-jacket' along which we all must now seek to travel. Rather, following some of the guiding principles outlined in Chapter 1 (see Box 1.1 and the subsequent discussion), it is, rather, to begin to assemble ways in which the integration and interdependence of questions of sustainability, security, economy, and health and well-being can be re-arranged in ways that create more medium- and long-term resilience and inclusivity in our food production and consumption systems. It is a paradigm of social and economic creativity and care, and critically one that also shines a strong light upon the existing unsustainable conditions of the 'worlds of food' we currently experience. It needs to be a paradigm that starts as much with the critical social and public priorities as with the economic; and it needs to be able to shape the latter through imagining and designing more effective and putative state-based interventions and actions.

The contributions to this volume have begun to tackle at least some of the key building blocks within this broad, critical and normative sustainability perspective. They have engaged with a range of theories and concepts to explore questions of food futures, food governance, the public realm and procurement, adaptive supply chains, biosecurity risks, animal welfare and political consumption, and the current implosion of the regulatory and geographical binaries between rural and urban spaces, all as new food spaces of hope and alterity.

It is clear from the sum of these contributions that current global and regional food systems are experiencing profound exogenous and endogenous pressures that are leading to far more complexity and uncertainty than many of the aggregated analyses of the food trends currently depict (see Morley *et al.*, Chapter 2, this volume). Indeed, while it will always be necessary to examine the broad shape of global food systems, as we describe in Chapter 2, and their potential futures and scenarios, this sort of genre needs to be clearly supplemented with far more in-depth, interconnected and diverse empirical and theoretically engaged research at multiple temporal and spatial scales. By focusing on syntheses of long-standing empirical research by the authors, this is partly what we have been attempting to do in the volume.

A major shift recognised, but perhaps somewhat unresolved in this volume, is associated with the declining dominance and increasing vulnerability of the European hybrid model of public–private regulation, which itself is a particular regulatory compromise between the increasingly neoliberalised state and the corporate food industry (especially retailers and major food processors). We developed an earlier in-depth analysis of this model in the previous research monograph emerging from the BRASS research programme in the early–mid 2000s, *The New Regulation and Governance and Food: Beyond the food crisis* (Marsden *et al*., 2010, p. 302). In conclusion to that volume we argued:

> Having traced the emergence of what can be described as a very complex but nevertheless robust and hybrid European model of public–private regulation, it is clear that this now faces new profound external challenges and vulnerabilities. ... First the model is becoming buffeted by the now well documented 'moral economy' surrounding food, whereby more consumers and NGOs persistently demand a more ethically and socially responsible set of food chains. Second, and more recently, we see the uncertainties, denial tendencies and highly contested debates surrounding resource constraints, especially as food scarcity and fuel costs drive up prices. Thus our model of food governance ... is not only now struggling with its own internal tensions and contradictory logics (such as continuing to increase the intensity of trade and movement in foods at the same time as scientifically reducing food quality and safety risks) ... it is increasingly located within a wider external set of global conditions which stress (the disproportionately richer) consumer moral economy, on the one hand, and the (disproportionately poorer) resource depleting carbon economy on the other. More than ever these changing macro and external conditions will challenge the coherent European model of food regulation; and they could throw it increasingly out of step with national and global governance priorities in the immediate future.

The contributions in this volume present a more contemporary and diverse picture of contingency, complexity, vulnerability and denial which has unfolded since the food crisis erupted in 2007–2008, and since we completed this earlier work on the complexity and dominance of the public–private model of food regulation. Clearly we see the boundaries of responsibility, legitimacy, regulation and governance shifting between the traditional binaries of the state, private and corporate sectors, and elements of civil society. The model depicted earlier is now clearly far more vulnerable. As neoliberalism hit its crisis just after the food crisis in 2007–2008, and governments set severe austerity policies in place as a reaction to this, further disruption of the assumptions and boundaries between what the State can and is willing to do began to change radically. Policy and resource management gaps began to open up, not just in the regulation and governance of food, but also more widely in the management of welfare and work. As Peck *et al.* (2011) and Larner (2012) argue more generally, neoliberalisation is still, even in crisis, an active and contested regulatory process, rather than just an event or an inanimate structural condition. Rather, its proneness to crisis and the spinning out of its risks create further innovative and oppositional social and political conditions for alternatives and what we call below 'short-circuiting'. As such, food welfare, poverty, ecological health and justice have risen as major structural and security concerns in recent years, compounding the endogenous problems of security, obesity and malnutrition already inherent in a malfunctioning food system (see Morley *et al.*, Chapter 2, this volume; Rayner and Lang, 2012).

Several new thematic and conceptual research avenues thus emerge during this more turbulent phase (post-2008) which we have been reporting on in the contributions here. These include: the power and legitimacy of food future's thinking and reporting, questions of neoliberal and potentially post-neoliberal (austerity) governance, the potential re-emergence of agricultural exceptionalism, public–private regulation of biosecurity and welfare, and the emergence of new urban–rural alliances for food sustainability. However, it is also important to reflect here upon some of our introductory arguments (see Marsden and Morley, Chapter 1, this volume) with regard to the historical and now contemporary positioning of food as a natural and, as such, 'awkward' dimension of capitalist and state development. It is important, we believe, to see the current sustainability and security crisis affecting food within the wider frame of global capitalist and technological developments, and to recognise that the study of food systems is in part an extremely valuable and critical heuristic for understanding wider and deeper formulations of the environmental and ecological crisis, or what Moore (2010) calls 'world-ecologies'.

In this sense we can argue that food is, metaphorically, a significant 'canary in the mine' in that its unsustainable-sustainable expressions (as, for instance, outlined in the chapters of this volume) give a clearer lens through which to see wider ecological and political economic shifts, tensions, and potential social and transformational opportunities. In a world where we still, after forty years of sustainability discourse, experience such a powerful set of forces of denial regarding the ecological crisis before us, it is important to use the paradigm of sustainable food – the ultimate metabolic and commodified materiality – as a heuristic to uncover the deeper and wider reasons for a lack of political and economic commitment and progress towards post-carbon transformatory change. We thus need, as a significant research endeavour, to continue to reassess the critical and dynamic status of food in its wider world-ecological ambience. Given its increasing centrality, once more, as an agent of progress for both human security and sustainable systems, as we introduced in Chapter 1, an important research vector concerns tracing its causes and effects with regard to the continuing and wider crisis of global warming and resource depletion.

Hot and cold foods: global heating and the beginning of the end of spatial and ecological fixes?

As the globe continues to warm, the race for carbon-based resources, as it currently stands, will only intensify and take further exploitative opportunities from the situation. At the time of writing, new data from the CryoSat satellite system has shown that sea ice levels in the Artic have decreased by four-fifths since 1980 while data from carbon dioxide monitors on the slopes of Mauna Loa in Hawaii show the second greatest annual leap in atmospheric CO_2 ever recorded (McKibben, 2013).The planet is heating up fifty times faster than at any point since human civilisation began according to temperature records dating back 11,000 years. Over the last thirty years vegetation zones in the Arctic have moved seven degrees latitude further north. Recent reports (Potsdam Institute for Climate Change Impact Research and Climate Analytics, 2012; National Research Council, 2013; Oreskes and Conway, 2013) confirm that because of the inertia in political and economic systems, it would now require an all-out effort to hold temperature increases below two degrees Celsius – the 'red line' that was drawn, but not acted upon at the Copenhagen Summit in 2009. Instead there is a constant push, fuelled by the denied but accepted emergence of 'peak oil', to drill, frack and mine for more oil, gas and coal. At the same time, as we saw in Chapter 4, regarding sustainable food supply chains, the agri-food corporate sector is still largely assuming relatively low transport costs and a rising demand for intensive meat production. Unlike in Germany, which is now supplying 22 per cent of its energy needs with renewable resources, and heading for more than 40 per cent within a decade, the fossil fuel industry has remained paramount in governing energy priorities in North America, Australasia and the UK. The World Bank, among others, regards increases of four degrees centigrade as devastating: in his preface Jim Yong Kim, the newly appointed President of the World Bank (Potsdam Institute, 2012), outlines the need to avoid the following consequences:

> [T]he inundation of coastal cities, increasing risks of food production leading to higher malnutrition rates, many dry regions becoming dryer; wet regions wetter, unprecedented heat waves in many regions, especially in the tropics, increasing frequency of high intensity tropical cyclones; and irreversible loss of bio-diversity, including coral reef systems.
>
> (quoted in McKibben, 2013, p. 59)

At the same time the fate of the contentious Keystone Pipeline, taking oil from Northern Alberta oil sands to the Gulf of the US, which if built would potentially lock in carbon-based burning for another forty years, is providing a particular problematic for a US president who has repeatedly identified climate change as a key priority.

Clearly then we are witnessing significant countertendencies to making sustainability transitions, as the carbon-based economy both intensifies its operations to vertically drill and to frack further into the earth's crust, while at the same time begins to – horizontally – experiment with plant and fibre-based bio-economic and refining options (see below). What this begins to indicate is the increasing interactions between the food–feed–energy–water–soils–land occupancy nexus and the balances and trade-offs that need to be struck between them. Critical here is the use and exploitation of increasingly limited land and fibre resources. The race for land-based food–feed–energy and carbon resources (new frontier spatial fixes) continues apace, despite risks of depletion and the limiting affects of climate change on weather patterns and therefore food production systems. As McMichael (2013, p. 48) recently argues with regard to the over 400 'land grabs' that have developed, overriding WTO trading rules, especially since 2008:

> [T]hese shifts involve (i) an emergent agro-security mercantilism by which certain states seek to guarantee access to food and biofuels via sponsoring direct acquisition of lands offshore, and (ii) a proliferation of governance mechanisms to justify and enable a new phase of land investments … Another way of capturing the seeming paradox is by characterising the land grab as pivoting on the dialectic of 're-territorialisation' via investment in offshore lands for agro-exporting of food, fuel and bio-economic products, and 'de-territorialisation' as host states surrender land and water from export to states defined (through market measures or policy) as food-dependent. The land grab opens up a new chapter in the redistribution of power across an increasingly multi-centric global food system with rising agro-export powers in middle-income countries (as expressed in the formation of the Group of Twenty (G20)).

As resources become tighter around the new land question, the management of these tensions and processes becomes a critical research issue, while price rises make the ability to create relatively 'cheap' surpluses all that more difficult.

As Moore (2010) postulates more generally, historically, capitalist phases of development are always predicated upon the creation of surplus from ecological regimes. These condition four key resource spheres: human and mechanical labour power, food, energy, and non-energy inputs such as metals, wood and fibres. The ecological and capitalist regime under neoliberalism (1980–2008) was based upon the continuous creation of cheap labour, energy and food, at least for a time. This logic is still driving the fracking bonanza in North America and Australia, therefore it is by no means dead. Yet the 2007/2008 food, financial, fiscal and fuel crisis was indeed an expression of the ecological limits of this cheap system of growth. Cheap food and other related resources, in particular, have always been connected and indispensable with the revival of global accumulation and growth, even in eras of financialisation (Moore, 2008, 2010).

In the latter half of the twentieth century, the radical decoupling of world prices and production costs created major new opportunities for the concentration and centralisation of capital in the agri-food sector. These processes would have been completely alien to Kautsky a century earlier as he analysed 'food futures' then and the distinctive and awkward nature of agricultural intensification and concentration during early industrialisation and urbanisation (Kautsky, 1899). By 2000, four corporations controlled 82 per cent of the beef packing in the USA, 75 per cent of hogs and sheep and half of the chicken industry (Greider, 2000). By 2008, five corporations controlled 90 per cent of the international grain trade, three countries produced 70 per cent of exported maize and thirty of the largest retailers controlled one third of the world's grocery trade (McMichael, 2009). In Europe, the Common Agricultural Policy and the liberal policies allowing the increasingly concentrated corporate retailers to source larger amounts of their groceries from the less developed world continued this bountiful sourcing of 'cheap food', under what we referred to in Chapter 1 as the post-productivist compromise.

What we have witnessed since, starting in 2003 and reaching a peak in 2008 is what Moore (2010) following Arrighi (1994) calls a major 'signal crisis' of neoliberalism as a (world) ecological crisis. While, as we have seen in many of the contributions to this volume, neoliberalisation lives on as Harvey (2009) calls, as a 'class-project', or as a mode of market-disciplinary regulatory restructuring (Brenner *et al.*, 2010), or more specifically here as an (anti-ecological and sustainable) late-carbon hegemonic project to continue to meet 'peak energy' demands, it is doing so increasingly upon borrowed time. As Moore (2010, p. 398) argues:

> [T]hese expressions of 'neoliberalization' are, in the final analysis, dependent upon the system's capacity to deliver cheap food, oil and inputs. Hence, signal crisis refers to the moment at which the ecological regime has reached its tipping point in the production of the relative surplus, the mass of use-values (appropriation) relative to the demands of world value production (capitalization). A terminal crisis still awaits.

In the past, and echoing the transitions management frameworks outlined by Geels (2002) and Spaargarren *et al.* (2012), a series of technological transitions and 'leaps forward' have usually ensured the growth of surplus and cheap food and energy supplies; even if, however, many of these have clearly relied upon imperial or post-colonial spatial fixes. These 'yield honeymoons' (such as with the 'green revolution' which was paralleled by mass proletarianisation) are, it can be argued, increasingly a thing of the past. Even the globalisation of the bioeconomy and bio-tech (European Commission, 2012, and see below) does not, however much it is being heralded as such, necessarily create a new sustainable platform or molecular frontier for cheap resources, with the rise in bio-tech plants yet to slow the progressive decline in global yield growth. The rapid development of the bioeconomy (see Kitchen and Marsden, 2011; Marsden, 2013) as a palliative to the under-production problems of agri-food and energy are, nevertheless, leading to the application of profound techniques that are attempting to control bio-physical nature at a cellular and genetic level. This is a transition from the 'formal to the real subsumption of nature to capital' according to Boyd *et al.* (2001) and Smith (2006), leading to potential 'third nature' corporate and neoliberalised solutions to the under-production of cheap food and energy problems. Yet this is yet to materialise, and may indeed be only a virtualism. Crop yields and productivity continue to at best plateau while corporate retailers and input suppliers such as fertiliser companies continue to find it more difficult to source cheap inputs. The land enclosure 'grabbing' development (Cooper, 2008; McMichael, 2013) could also be seen as an attempt to 'solve' and spatially fix the current scarcity of external frontiers. But this seems unable to assuage a growing consensus among scholars and policy bodies that the times of 'cheap food' and 'cheap oil' are finished (OECD, 2009, 2011).

Climate driven reductions in crop land of the order of 8–20 per cent by mid century, mounting water scarcity concerns, the proliferation of invasive species, rising biological resistance to pesticides and herbicides, escalating land competition for arable versus agri-fuels, and an absolute decline in productive land area of the order of 12 per cent of the planet, all are dovetailing with the aforementioned rises in global heating and its associated and entrenched politics of denial. As such the neoliberalising regulatory model of world ecology is fast running out of spatial and ecological 'fixes'. The current conjuncture does thus have all the signs of Arrighi's signal crisis; a crisis in which the current world ecology is quickly running out of cheap options on which to secure its necessary resources and capitalisation processes.

The point here is not to simply be reactively apocalyptic about the particular unsustainable conjuncture at which world ecology now finds itself, but to recognise how the social and political realisation of these conditions can and could act to stimulate a creative platform for new innovative pathways for sustainability based upon a post-neoliberal model of development, governance and regulation (Brenner *et al.*, 2010). This could involve new governance models based upon public priorities, citizen sovereignty and 'progressive protectionism' (see below). The question is: where and how will this occur?

In Chapter 1 of this volume we started by explaining the long-running problems and awkward conditions that agri-food developments present for wider capitalist (urban and rural) development and accumulation. The interconnected problems of sustainability and food security were partly 'solved' in the late nineteenth century by strictly defining and functionally regulating the city and the countryside and applying fertilisers for more intensive agricultural regions. During much of the twentieth century this compromise was extended through technological yield promoters and state intervention to further stimulate intensive systems of production for mass consumption. Between 1980 and 2000 these systems, now backed by a more explicit neoliberal regulatory system, offered the lowest world market prices for food in history, accompanied by a massive expansion of cheap working-class labour and relatively cheap fuels. Under these conditions, barring a few severe food safety scares in the 1980s and 1990s, cheap food, and indeed more exotic cheap food, was largely taken for granted by those living in advanced societies. This further fuelled neoliberal governments to encourage over-consumption of luxury and leisure goods as 'trophies' of success and progress. This was indeed the 'golden age' for Western consumers, who could begin to protect their own environments at the same time as consuming a greater proportion of world food goods at a cheap rate. As such, excessive and unsustainable consumption patterns were legitimised and reinforced, in turn, a strong political basis for continued neoliberalisation on the one hand and corporate- (retailer) led expansion on the other.

As the contributions to this volume have indicated, the period 2007–2008, signalled the end of this period of assumed stability, bounty and plenty. The combined financial, fuel, food and fiscal crises that have ensued since are indeed all part of the same coin. Rising production costs of labour, food, energy and non-energy materials – the long-term lifeblood of uneven capitalist accumulation – are all facing bio-physical barriers and limits, however much these may still be denied. Moreover, this is still being reflected in financial markets, corporate strategies and state responses to the crisis. Indeed, the 'time–space compression central to the accumulation of capital both depends upon and drives ever faster the time–space compression of bio-physical natures'. The 'political ecology of Nature is both an "opportunity" and "obstacle" – an enabler of, and hindrance to, capital accumulation – in successive ecological regimes' (Moore, 2010, p. 407).

So it would seem that while it may be timely to build a new sustainability paradigm for food it is also quite fraught with the entwined difficulties of agri-food's particular and clearly interdependent niche in the current bio-physical and human conjunctures that now face us. Clearly, as this discussion demonstrates, we cannot and should not separate agri-food developments from the wider resource governance and ecological crises that currently confront us. Indeed it is essential that agri-food scholars fully engage with these wider 'world-ecology' debates as the current conditions, with all their contradictions and contingencies, ensue. Given this context, it is important to begin to conceptually build new frameworks and insights about how creative and innovative sustainability solutions and adaptations can be embedded and constructed.

'Re-placing' neoliberalism: food as a means and agent of change for wider sustainable community development

Agri-food scholarship, especially since the beginnings of the food and related crises of 2008 has begun to locate itself very much in the heartland of wider social science debates concerning not only environmental and ecological crisis but also the shifting boundaries and expectations between the multilevel state, the private corporate and non-corporate sectors, and civil society. The combinative food, financial, fiscal and fuel crisis, and its particular state reactions of policies of austerity, are creating a new research and policy vector for agri-food and agri-food studies. In particular, as we outlined in Chapter 1 of this volume, this centres around how to reintegrate questions of food security with those of sustainability; and how the state, private and public sectors can create the conditions for new sustainable development pathways in a time of vulnerability and crisis in existing systems. In this sense the critical study of agri-food as part of the wider resource nexus outlined above becomes a key feature of understanding and enacting new societal development paths. After the modernisation paradigm of the twentieth century, as we have seen, new multiple modernities are needed that relocate food as a key vehicle in societal development, just as it was in the early industrialisation and urbanisation modernities of late nineteenth and early twentieth century Europe.

A sustainability paradigm for agri-food then needs to begin to explore these new vectors and opportunities in contributing to multiple sustainable development pathways. These new conceptions, as we begin to see in the contributions in this volume, transcend many of the traditional conceptual binaries that are still dominant in the literature: urban–rural, nature–society, conventional–alternative, intensive–extensive, agency–structure, macro–micro. We need new hybrid conceptualisations that begin to capture both what is and what might be as we begin to attempt to reintegrate sustainability and security in the food realm. Clearly, the question of new (post-neoliberalised) state and governance mechanisms and innovations is critical. We see in Lee's chapter (Chapter 3) and in Morgan and Morley's chapter (Chapter 4) how either at the European or at national government levels the state now has to grapple with integrating sustainability and security issues from a base of competing and often opposing policies (such as agricultural exceptionalism and market liberalisation). Also, in the management of diverse biosecurity risks (Enticott, Chapter 6) and growing but uneven demands for animal welfare as part of the sustainability/security mix (Miele and Lever, Chapter 7), the extant state, at multiple levels is currently limited to rather traditional market-based mechanisms of regulation and governance. Critically, these are based upon the continued control of risks rather than supporting a new infrastructure of transition.

Here lies a major weakness of current food governance. It is not yet fit for purpose in aiding and fostering the food security and sustainability transitions and adaptations that are needed (see Lee, Chapter 3). These are now being witnessed empirically in the rise of cities and towns as new policy actors (see Sonnino and Spayde, Chapter 9), and in the contradictions and inertias associated with the continued policy binaries that exist between the planning and management of urban and rural places, documented in Franklin and Morgan's chapter (Chapter 8). Nevertheless, these contributions do indicate that policy 'spaces' are opening up for novel civic, private and cooperative ventures to take place and become embedded in certain places, as part of a post-neoliberalist agenda. As Colin Hines argues (2013), it is timely to consider new ways in which the state can foster place-based cooperative policies and experiments that protect and re-build Europe's local economies. This is to advocate a new type of 'progressive protectionism' and place-based sufficiency that clearly lies in opposition with the still dominant discourse and policies of neoliberal 'free-market' trade and comparative advantage. These place-based systems of progressive protectionism and resilience need not be seen in opposition to the growth of fair and ethical trading, as they should strike a new balance between fostering the local and regional eco-economy (of which food is one central element) with embedded and sustainable trading relations, locally, nationally and internationally. The growth of trans-local connections in food initiatives is one important development in this regard; demonstrating that there is not necessarily a contradiction between place-based progressive protectionism and international trade, networking and knowledge flows.

In this sense it is important for agri-food scholarship to consider more sustainable and secure *places and spaces of possibility and experimentation* whereby, through new collaborative public and private actions, sustainability and security can be re-connected. Over the past decade much of agri-food scholarship has documented the rise of the alternative food movement, very much in something of a conceptual vacuum. This has largely failed to address questions of potential convergence and scaling out, or the extent to which these alternatives provide a real basis for progressive political contestation and development or sustained post-neoliberalised and post-carbonised transition. In many ways, this absence of this critical engagement, and a focus upon the micro and internal dynamics of alternative food initiatives has, with hindsight, created something of a conceptual local 'trap' for alternative and critical agri-food scholarship. It has been somewhat easy for critics to render this scholarship as inherently marginal and ideographic, and to elide local with bland notions of sustainability.

Alternative and progressive agri-food scholars need to shake off this conceptual trap with which some may wish to encircle and marginalise them. The evidence in this collection is that 'alternative agri-food' is far more significant (both potentially and existentially) than conventional aggregated metrics of scale, volume and geography would suggest. This takes Gibson-Graham's (2006) and Friedmann's (2007) notions of multiple deconstructed place-based economies to new levels. And it poses the question of how governments and the ways we organise our economies can give more space and support to the planning of public and civic eco-economies. These indeed start with the social and public priorities of food security and sustainability – through community development, building social capital and skills, sustainable nutritional feeding programmes, sustainable production, linking food provenance and the provision of energy, housing and transport.

The distinguishing feature of eco-economic development in agri-food, as in renewable energy or related sustainability arenas, is that they indeed start with the social priorities and agendas. There is no doubt that across Europe and parts of North America, and indeed more extensively in countries such as Brazil, we are beginning to witness the growth of interconnected eco-economies that are significantly 'short-circuiting' conventionally regulated food supply chains (see van der Ploeg and Marsden, 2008; Horlings and Marsden, 2010). These new 'webs', hubs and alliances are developing on a place-based basis and often creating external trading relationships between urban and rural places. They are being financed by new innovative partnerships between banks, NGOs and private sector firms, such as through Regional Accounts, which are money and credit making systems for sustainability projects (for example the Het Groene Woud system in the Netherlands, which has created a spending budget of 200,000 euros a year) (Rural Alliances, 2013).

We are thus witnessing a growth in the complexity and density of developments in the agri-food sector that necessitates a wider and less binary conceptual approach to 'alternative food movements'. Rather, these need to be located within a wider vision and more variable vector of bio-economic and eco-economic developments that celebrate the diversity of this complexity. A challenge to it, however, is the rather narrow techno-science notions of the bioeconomy, whereby it is assumed that the sustainability and food security conundrum can be 'fixed' by specific technocratic notions of 'sustainable intensification' (see Morley *et al.*, Chapter 2). Agri-food scholars need to constructively and critically engage with these more restricted and technocratic 'solutions' to the crisis. We need to see this as part of a contested process of public/private knowledge creation and political framing that involves differential levels of control over human–natural relations. Concepts of the bioeconomy and its relationships to achieving 'sustainable intensification' are now a powerful scientific and political framing device. The EU and the OECD have recognised this in various policy statements (see OECD, 2009, 2011; EC, 2012). In rational terms the OECD (2009, p. 9) defines the bioeconomy as:

> part of the economic activities which captures the latent value in biological processes and renewable bio-resources to produce improved health and sustainable growth and development ... the bio-based economy, deals more narrowly with industrial applications: it is an economy that uses renewable bio-resources and eco-industrial clusters to produce sustainable bio-products, jobs and income.

The EU (2012) estimates that the bioeconomy already has a turnover of 2 trillion euros and employs more than 22 million, or 9 per cent of total European employment. It includes exploiting the intersections between the traditional sectors of agriculture, forestry, fisheries, food, pulp and paper production, as well as the chemical, bio-technological and energy industries. One can see here a conceptual and framing alignment between notions of sustainable intensification in food and the wider vector of bio-economic development that is linked to notions of 'sustainable growth'. In the food sector this is leading to a wider array and experimentation with a range of functional foods (including in vitro, health enhancing, lower calorie and nano-foods) as well as proposals for 'closed loop' mega farms (see Horlings and Marsden, 2010; Kitchen and Marsden, 2011). Clearly the level of public acceptability, legitimacy and real market potential of these new bio-economical food products is an open question; as indeed is how government and regulatory authorities are going to handle their risks. Whatever, it is clear that one significant technocratic and scientific reaction to both the carbon and the food crisis is to speed up the frontier of molecularisation, and indeed to do so, in contrast to many alternative movements, by privatising those new knowledges and patents. One important area of research here becomes exploring the different balances between privatised and public knowledges of food innovations and developments. Monsanto's recent successful legal battles against 'rogue' farmers who have dared to attempt to re-seed their GM-base products, is a clear example of how private rights and patents can be used to continue to commodify seed varieties, and to create further producer dependence on these seed commodities (Deibel, 2013).

It is clear that in some parts of the world, and partly depending upon the strength and balance of public–private regulation and governance, particular types of bio-economic experimentation are likely to become a dominant feature of 'sustainable intensification'. However, in doing so they are likely to create social and political opposition, the strength of which is so far unforeseen. This variable oppositional status is and will continue to be a major vehicle for the further development of alternative eco-economical developments, driven by different combinations of political producers and consumers (see Miele and Lever, Chapter 7, this volume). In Europe, we see new geographies of political consumption holding different degrees of power in different member states and regions; and the EU residing over a pan-European contradiction of maintaining and upholding a moratorium on the importation of GM products, while explicitly stimulating the research and development associated with the bioeconomy. In the UK the main corporate retailers, while attempting to espouse more local and national sourcing of meats following the recent horsemeat contamination, are also admitting to its customers that they can no longer guarantee that their chicken products will be GM feed free. This is owing to an alleged shortage of non-GM soya feed in global markets. Hence, we see increasing contradictory trends and actions as a clear feature of the new 'sustainable foodscape', with different and competing interests developing and articulating their own 'strong' or 'weak' framings of sustainable development and intensification. These variable framings should not divert scholars from critically exploring 'real' forms of food sustainability. For, as we have argued throughout this volume, the twin societal lenses of potentially integrating sustainability and security are critical features of a more engaged and progressive agri-food scholarship.

'Re-placing' neoliberalism will only occur in some places at the potential expense of others, for as we have witnessed in the historical development of agri-food in capitalist development, it is both a uniquely distinctive metabolic and politicised process, which continually re-recreates places and spaces through a constant process of uneven development. The sustainability paradigm needs to recognise and begin to reclaim control, ameliorate and shape these processes of uneven development as the world continues to become more vulnerable for larger segments of its population. At the same time it is clear that there is growing public concern for the need to link sustainability and food security issues and policies among larger sections of consumers and civil society. Health, social justice and food sovereignty are amalgamating with local economic and ecological concerns to create more powerful alliances and partnerships in and across rural and urban spaces. The very fact that the majority of the world's population will be urban dwellers over the rest of this century only reaffirms the necessary synergies and interfaces needed between urban consumer energy and food demands and their reliance upon their wider resource hinterlands. No single technological or spatial fix is going to solve these new 'food equations'. They will require at the very least 'necessary' and reflexive governance systems that foster sustainable solutions that create resilience out of potential vulnerability.

The current dominant political discourse regarding increasing production and productivity of food supply needs to be matched in equal fashion with analysis of how more people gain access to land, food and water resources. These are essentially governance concerns rather than singularly technological or economic. They will require new innovations in governance systems to allow access to increasingly scarce resources. Food and agri-food studies will need to be at the forefront of designing and critically assessing new reflexive governance mechanisms and processes that can accommodate these new consumer and producer demands. Engaging with policy and political developments, with a wider vector of civil, public and private agri-food actors and stakeholders will be fundamental, such that the co-production of sustainable knowledge can inform the public realm. Moreover, political leaders and their officers will need to be more receptive to the findings of scholarship in the agri-food realm, as we are beginning to witness in some of the food council and city strategy processes across the world. In turn, as researchers we will need to make our knowledges more 'open source', accessible and adaptable to institutional adoption and to be wary of diluting processes of co-option in the name of many forms of 'weak sustainability'.

In conclusion here we have attempted to begin to trace aspects of the rich research agenda for a more engaged and theoretically challenging agri-food studies during a period that has witnessed a 'signal' transformatory moment (to use Arrighi's phraseology). Amidst this more turbulent and contingent context it is important, as we have outlined here, for scholars to think through rigorously *what might be* as well, of course, as *what is*. In addition, it will be increasingly necessary to consider *how* and not just *whether* current conditions should change and in what direction. In this context there may be multiple modernities and multiple and layered multiple economies, as a more urban-based society redefines its relationships with its natural hinterlands. Agri-food and agri-food relations will continue to become a major fulcrum for enacting the sustainable transformations needed in different national, local and regional settings.

In this context, the 'local' becomes a creative place not just of resistance but for doing things differently, in being socially and economically creative and connected in and across different spaces. Local development may be the place to start the social and interconnected mobilisations necessary for re-integrating sustainability and food security. In focusing on the local we will need to incorporate the global and external interfaces that partly constitute these local conditions. Under such conditions we will also need to devolve centralised power and regulation such that it provides opportunities for local and place-based creativity and social innovation to take hold. This will be a constant challenge for existing governance and regulatory bodies and will demand more nuanced functional distinctions to be made as to what policies need to be shaped and implemented and at what spatial scales of jurisdiction. Food politics and the broader politics of resource scarcity and consumption will be a powerful influence in driving these necessary governance changes in the future as consumers and producers become more conscious of the provenance and indeed social and ecological value of the food and wider land-based goods they create, purchase and procure.

References

Arrighi, G. (1994) *The Long Twentieth Century*, London: Verso.

Boyd, W., Prudham, W. and Schurman, R.A. (2001) 'Industrial dynamics and the problem of nature', *Society and Natural Resources*, vol. 14, pp. 555–570.

Brenner, N., Peck, J. and Theordore, N. (2010) 'After neoliberalisation?', *Globalisations*, vol. 7, no. 3, pp. 327–345.

Cooper, M. (2008) *Life as Surplus*, Seattle, WA: University of Washington Press.

Deibel, E. (2013) 'Open variety rights: Rethinking the commodification of plants', *Journal of Agrarian Change*, vol. 13, no. 2, pp. 282–389.

European Commission (2012) *Innovating for Sustainable Growth: A bio-economy for Europe*, B-1049, Brussels: European Commission.

Friedmann, H. (2007) 'Scaling-up: Bringing public institutions and food service corporations into a project for a local sustainable food system in Toronto', *Agriculture and Human Values*, vol. 24, pp. 389–398.

Geels, F.W. (2002) 'Technological transitions as evolutionary reconfiguration processes: A multi-level perspective and a case study', *Research Policy*, vol. 31, no. 8/9, pp. 1257–1274.

Gibson-Graham, J.K. (2006) *A Post-capitalist Politics*, Minneapolis, MN: University of Minnesota Press.

Greider, W. (2000) 'The last farm crisis', *The Nation*, 20 November.

Harvey, D. (2005) *A Brief History of Neo-liberalism*, Oxford: Oxford University Press.

Harvey, D. (2009) 'Is this really the end of Neo-liberalism?', *Counterpunch*, 13–15 March.

Hines, C. (2013) 'Europe and the fight for fair wages', *The Guardian*, 20 May.

Horlings, I. and Marsden, T.K. (2010) 'Pathways for sustainable development of European rural regions: Eco-economical strategies and new urban–rural relations', ESRC BRASS Working paper No. 55, Cardiff: Cardiff University Press.

Kautsky, K. (1899/1988) *The Agrarian Question (Vol. 2)*, London: Zwan Publications.

Kitchen, L. and Marsden, T.K. (2011) 'Constructing sustainable communities: A theoretical exploration of the bio-economy and eco-economy paradigms', *Local Environment*, vol. 16, no. 8, pp. 753–769.

Kuhn, T.S. (1962) *The Structure of Scientific Revolutions*, Chicago, IL: University of Chicago Press.

Larner, W. (2012) *C-Change? Geographies of Crisis: Dialogues in human geography*, London: Sage.

Marsden, T.K. (2013) 'Sustainable place-making for sustainability science: The contested case of agri-food and urban–rural relations', *Sustainability Science*, vol. 8, no. 2, pp. 213–226.

Marsden, T.K., Lee, R., Flynn, A.C. and Thankappan, S. (2010) *The New Regulation and Governance of Food: Beyond the food crisis*, London: Routledge.

McKibben, W. (2013) 'Some like it hot! Review', *New York Review of Books*, 9 May, pp. 59–60.

McMichael, P. (2009) 'The world food crisis in historical perspective', *Monthly Review*, vol. 61, no. 3, web edition.

McMichael, P. (2013) 'Land grabbing as security mercantilism in international relations', *Globalisations*, vol. 10, no. 1, pp. 47–64.

Moore, J. (2008) 'Ecological crises and the agrarian question in world-historical perspective', *Monthly Review*, vol. 60, no. 6, pp. 54–63.

Moore, J. (2010) 'The end of the road? Agricultural revolutions in the capitalist world-ecology, 1450–2010', *Journal of Agrarian Change*, vol. 10, no. 3, pp. 389–413.

National Research Council (2013) *Climate and Social Stress: Implications for security analysis.* J.D. Steinbruner, P.C. Stern and J.L. Husbands (eds), Washington, DC: National Research Council.

OECD (2009) *The Bio-economy to 2030: Designing a policy agenda. International futures programme*, Paris: OECD.

OECD (2011) *A Green Growth Strategy for Agriculture: Preliminary report*, Paris: OECD.

Oreskes, N. and Conway E.M. (2013) The collapse of Western civilisation: A view from the future, *Daedalus*, Winter 2013.

Peck, J., Theordore, R. and Brenner, N. (2011) 'Neo-liberalism and its malcontents', *Antipode*, vol. 41, pp. 94–116.

Potsdam Institute for Climate Change Impact Research and Climate Analytics (2012) *Turn Down the Heat: Why a 4C warmer world must be avoided*, Potsdam: Potsdam Institute for Climate Change Impact Research and Climate Analytics, November.

Rayner, G. and Lang, T. (2012) *Ecological Public Health: Reshaping the conditions for good health*, Abingdon and New York: Routledge.

Rural Alliances (2013) 'Case study – Financial engineering: Partnership and landscape fund for landscape development in industrialised areas', Briefing note, *Rural Alliances*, June 2013.

Smith, N. (2006) 'Nature as accumulation strategy', *The Socialist Register*, vol. 43, pp. 16–36.

Spaargarren, G., Oostervier, P. and Loeber, A. (2012) *Food Practices in Transition*, London: Routledge.

van der Ploeg, J.D. and Marsden, T.K. (eds) (2008) *Unfolding Webs: Theories of rural and regional development*, Assen, Netherlands: Royal van Gorcum Press.

Index

Page numbers in *italics* refers to a table/ figure

Achieving Food Security in the Face of Climate Change 50, 53
Adhikari, J. 187
affluence: and nutrition transition 36–7, *36*
Africa: population of 34
agri-environmental schemes 7
agri-food corporatism 8
agri-food sector 216; concentration and centralisation of capital in 211
agri-industrial model 3, 4, 6, 169
agricultural exceptionalism 6, 73, 78, 208; and EU 62–8
agricultural population 35, *35*
Agricultural Wages Board 64
agriculture: animal welfare and sustainability in 143–4; energy consumption 38, *39*; and environment 77; foreign direct investment 46, *46*; greenhouse gas emissions 41, *41*; impact of climate change on 42, *42*; inefficiency in 51; investment in 46–7, *46*; land use 43, *43*; multifunctional nature of 168–9; and non-CO_2 gases *41*, 42; organic 47, *47*, 115; pathway to 2050 54, *54*; pillars of sustainable 54–5; production and yields 43–5, *44*; R&D funding 47; regulation of workers' remuneration 64; reliance on nitrogen fertilizers 39, *40*; state support 45, *45*; storage of commodities 44; transgenic crops 45, *45*; use of fossil fuels 38–9; water use 38
Agriculture Act (1947) 5, 63, 172
Alemanno, A. 68
alternative food hubs 19
alternative food movements 20, 21, 215, 216
Alternative Food Networks 2, 11, 19
alternative food systems 110–11
Amsterdam Treaty (1997) 143
animal farming: regulation of by law 144–5
Animal Health Board (AHB) (New Zealand) 129, 131, 132, 134, 136, 138–9
Animal Health and Welfare Board for England (AHWBE) 135
animal welfare 143–58, 168; as an element of agricultural sustainability 143–4; and EU Animal Welfare Strategy 158, 160–1; and European Animal Welfare standard 158–9, 160–1; France 147, 153; growing public concern with 144; Hungary 152–4, 157; institutional setting models 145–9; Italy 147, 150–1; and labelling 144, 147, 149, 150, 151, 152, 156, 157, 160; marketization and political consumption of 156–7, 158–61; as a multidimensional concept 144–6, 158; national assurance schemes 148, 149–50, 151–2; national differences 158; Netherlands 148, 150, 155, 159; no issue model 145, 146; Norway 148, 152, 155, 156; and regulating animal farming by law 144–5; and structural variations among animal supply chains 153–8; supermarket model 145, 148–53; Sweden 147–8, 151–2, 155, 156; terroir model 145, 147; welfare state model 145, 147–8
Animal Welfare Strategy *see* EU (European Union)
animal-based products, consumption of 36–7, *36*
Ansell, C. 71
Arctic sea levels 209
Aristotle 94

Arrighi, G. 211
assurance schemes: and animal welfare 148, 149–50, 151–2
Assured Food Standards (AFS) 149
austerity policies 88, 100, 208
Australia: and Bovine Tuberculosis 137–8

badger culling 125, 128, 133, 135–7, 138
Barker, K. 125–6
Barlett, P.F. 90
Best Value 87
biodiesel 40, *40*
biodiversity 13, 15, 52, 65, 107, 133, 193
bioeconomy 12, 122, 123, 137, 139, 212, 216–17; definition 216; estimated turnover 217
bioethanol 40
biofuels 40, 76
Biofuels Directive (2003) 76
biosecure citizenship 125–6, 132, 136, 139
biosecurity 122–39; and animal disease prevention 124; Bovine Tuberculosis (bTB) case 128–37; as a local solution to local agricultural problems 124–5; national scale 125–6; and neoliberalism 127, 138–9; newspaper articles mentioning 124; spatial scales of 123–8; territorial and international responsibilities 126–8
bird flu 122
Blay-Palmer, A. 197
Blokhuis, H. 158
Bloom, J.D. 105
bluetongue virus 122
Bohle, H.G. 187
bovine spongiform encephalopathy *see* BSE
Bovine Tuberculosis: Australia 137–8; Britain 128–9, *129*, 133–7, 138; New Zealand 128–33, *129*, 134, 136, 137–8
BRASS research programme 25, 78, 207
Braun, B. 126
Brazil 19, 45; overweight individuals 13; schools meal provision 92
Brighton 192
Bristol 198; food policy council 189, 190
Brown, K. 175, 182
BSE (bovine spongiform encephalopathy) 7, 68, 69, 155
bulk farming strategy 170
Buller, H. 76–7, 143, 160
business perspectives: and food future reports 53–6
buyer–supplier relationships 105–6

calorie energy consumption 43
Cameron, David 76
CAP (Common Agricultural Policy) 63, 64, 65, 73, 77, 167, 170, 173, 211; alterations to Pillar I 168; Pillar II 168; proposed strategic aims 170; reform of 66–7, 71–2, 168
Capital Growth initiative 175–6
capitalism 6
carbon dioxide emissions 209
carbonism 20
Carrefour 104
catering: evolution of public sector 85–9
charters 16
Chatham House report 78, 116
chickens, organic 111–12
China 9; meat consumption 36; overweight individuals 13; R&D funding for agriculture 47
cities: emergence of as food policy actors 187–9; and Food Policy Councils *see* FPCs; reconnecting with the countryside 11–14, 23; spatial partitioning of countryside and 5, 7; and Urban Food Strategies *see* UFSs; *see also* urban strategies
citizenship, biosecure 125–6, 132, 136, 139
class structures, agrarian 6
climate change 3, 9, 41–2, 117, 173, 209; agriculture and greenhouse gas emissions 41, *41*; and futures reports 53; global warming 209; impact of on agriculture 42, *42*; and water availability 38
Common Agricultural Policy *see* CAP
community food growing initiatives 56, 166, 172, 174–83; and access to land 175–7, 181–2; community groups becoming freehold landowners themselves 182; farmer entrepreneurialism and rural community landshare 179–81; and social relationships between landowners and groups of land users 181–2; Stroudco community food hub 177–9
community food hubs 177
community supported agriculture (CSA) 174
compact city development 172
Compass 91, 97
compulsory competitive tendering (CCT) 87
Conservative–Liberal coalition 16–17

Consultative Group on International Agricultural Research (CGIAR) 53
Coombs, R. 109
corporatism 11–12, 63–4, 73
corporatist governance: features of 17; versus reflexive governance 16–18
Council for Rural Communities 17
countryside: reconnecting with the city 11–14, 23; spatial partitioning of city and 5, 7
Cox, A. 106
Crush, J.S. 188
CSA (community supported agriculture) 174

Daugbjerg, C. 67
Debio 152
Delaware Valley Regional Planning Commission (Philadelphia) 192
Department of Environment, Food and Rural Affairs (Defra) 76, 133–4, 136, 137
deterriorialization 126–7
developing countries 9, 12–13, 66, 187; animal-based consumption 36; effect of increased prices 116
disease, animal 122; Bovine Tuberculosis 125, 128–37; ways of preventing 124; *see also* biosecurity
Doha Round 66, 67, 71
Douglas, Roger 130

eco-economy 13, 15, 215, 216
Ecoli crisis (Wales) (2005) 100
ecological footprinting 20
ecological model: and school meals 87–8
Education Act (1944) 86; (1980) 86
Education (Provision of Meals) Act (1906) 86
EFSA (European Food Safety Authority) 70–1, 73, 74, 75
energy 38–40; consumption of by agriculture 38, *39*
Enterprise and Regulatory Reform Bill 64
Enticott, G. 136
environment–economy integration 15
environmental conservation: and UFSs 194
environmental impacts of food system 114–16
environmental protection: and European food governance 76–7
environmental sustainability: and futures reports 52–3
Environmental Sustainability Vision Towards 2030 50, 53
equestrian livery yards 180–1
equity challenge 56
ethanol production *40*
ethic of care 94
ethical reasoning 94
EU (European Union) 8–9; and agricultural exceptionalism 62–8; Animal Welfare Strategy 158–9, 160–1; enlargement 64; public procurement regulations 85
EU LEADER 22
European Animal Welfare standard 158
European Commission 71, 77, 158, 170
European Court of Justice 74
European food governance 62–79; and agricultural exceptionalism 62–8, 73, 78; and CAP 63, 64, 65, 66–7, 71–2; and disputation 71–2; disputes and reasons for 62; energy and biofuels 76; and environmental protection 76–7; export subsidies 65; food and market liberalization 68–73, 78, 79; and food safety 68–9, 70, 71; and GATT 65; and GM crops 70–1, 72, 74; and health 77; market liberalization 62–3; problems for 75–6; science-based approach 73–6; towards integrated 76–8
European Food Safety Authority *see* EFSA
European Parliament 69, 72
Evans, A. 160
export subsidies 65

Fainstein, S. 197
fair trade 12, 47, 110, 111, 159
family farming 4, 6
FAO (Food and Agricultural Organization) 13, 15–16, 19, 31, 43, 46, 49, 186–7; Consumer Price Index for Food 32; food price indices 32, *32*; 'One Health' campaign 122
Farm Bill (2008) (US) 89
farmer entrepreneurialism: and multi-functionality 167–71; and rural community landshare 179–81
farmers: and community food growing initiatives 177; relations with retailers 107
farmers' markets 48, 111, 174; in Stroud 178–9
Fauna Society 146
Federation of City Farms and Community Gardens (FCFCG) 176
fertilisers 4, 213; nitrogen 39, *40*

financial crisis (2008) 1, 9, 30, 88
Finland 20
FMD (foot and mouth) 122, 124, 125, 133–4, 155
Fodor, E. 146
food: as part of world-ecology 206–9
food access 34, 48; inequality of 43
Food and Agricultural Organization *see* FAO
Food and Agriculture: the Future of Sustainability 50, 55
food crisis (2007/2008) 2, 3, 23, 24, 99, 116, 207
food futures/food futures reports 30–61, 207; alternative 56; business perspectives 53–6; and climate change 53; and environmental sustainability 52–3; food production and security 51–2; implications for building sustainable 56–8
Food for Life Partnership (FFLP) programme 87, 91, 92, 96
Food Policy Councils (FPCs) 186, 188, 189–92, 193, 195–6, 197; definition and role 189; educational role 191; encouraging change 190–1; engaging with stakeholders 191; functions 190–1; giving advice 190; reliance on involvement of civil society 190; research framework for sustainable 201–2
food prices 44; factors impacting on 32–3; impact of high 33; inflation and growth of 1, 33–4, *33*, 37, *37*, 116
food procurement: in public sector 85, 89–93, 100
food production: futures reports 51–2; historic growth rates 51
food safety 68–9, 73, 75, 79, 103, 105, 168; and European food governance 68–9, 70, 71; regulations 7, 62; *see also* EFSA (European Food Safety Authority)
food security 3–7, 9, 14, 169; crisis in 10, 11; and food sustainability 15–16, 20–1, 23–4; futures reports 51–2; pillars of sustainable 54–5; urban strategies for 186–202
food sovereignty movement 12
Food Standards Agency (UK) 17, 74
Food Strategy (2020) 17
food supply chains 103–18; alternative food systems 110–11; buyer–supplier relationship 105–6; and food systems 108–10; and globalisation 104–8; impact of on environment 115–17; local 110; organic food 111–14; own label brands 105; retailer–farmer relations 107; and retailers 104; supermarkets and environmental supply chain management 106–7
food system: challenges to be met 56; environmental impacts 114–16; and food supply chains 108–10; trends and the current state of 32–48
food waste *see* waste, food
FoodDrink Europe 53
foodscape 12
Foot and Mouth *see* FMD
foreign direct investment: and agriculture 46, *46*
Foresight report 57
fossil fuels 38–9
FPCs *see* food policy councils
fracking 210–11
France: agricultural sector 153; animal welfare 147, 153
Francis, M. 106
Franklin, A. 136
Frayne, G.B. 188
Freedom Food 149–50
Freidberg, Susanne 148–9
Friedmann, H. 215
future *see* food futures
Future of Food and Farming, The 50, 51, 52, 53
futurity 15

Geels, Frank 109–10
General Agreement on Tariffs and Trade (GATT) 65
genetically modified crops *see* GM foods
Ghana 92
Gibson-Graham, J.K. 215
GildePrior 152
Gill, S.E. 173
global warming 209
global–local relations 24
globalisation 116, 212; and food supply chains 104–8
GM (genetically modified) foods 45, *45*, 52, 70, 74, 170; European food governance 72
Godenhjelm, S. 21
Godfray, H.C.J. 51, 52
'Good Food for London' report 191
good food system 13, 193, 194
Grant, W. 133
green building design 173

Green Revolution 30, 211
greenhouse gas emissions: agriculture 41, *41*
greenspace: and urban planning 173
greenwash 56
Greenwich Co-operative Development Agency (GCDA) 191
Groceries Code Adjudicator 106
Groceries Supply Code of Practice 106
Guthman, J. 114

Habitats Directive (1992) 65
Hall, J. 106
health 12–13; and European food governance 77; and food 18; *see also* obesity
Heffernan, W.D. 6, 111
Hendrickson, M.K. 111
Hendrix College (Arkansas) 90
Hines, Colin 215
Hinrichs, C.C. 105, 174
Holocaust 94
Home Grown School Feeding programme 92
horsemeat scandal 78, 79, 99, 107, 217
How to Feed the World in 2050 50, 51, 52
Howard, P. 113
Hungary: agricultural sector 153–4; animal welfare 146, 152–4, 157

IAASTD report 23
Iceland 107
IFPRI 44
imperial food regime 5
income disparities 20–1
income growth: and nutrition transition 36
India: R&D funding for agriculture 47
inequalities, spatial and social 20–1
Institute for the Future 56
integrated supply chain management 105
intensive food regime 4, 5, 7, 9
Intergovernmental Panel on Climate Change 38
international food order 6
Italy: animal production 154; animal welfare 147, 150–1

Jacobs, M. 14
Jaggard, K.W. 52
Jamieson, Dr Sam 130, 132
just-in-time system 8, 22, 48

Kansas City Food Circle 111
Kautsky, K. 4, 211
Keystone Pipeline 210
Kirwan 110, 111
Kjærnes, U. 145, 147
knowledge deficit 96
Koos, S. 159, 161
KrAv 152
Kroger 105
Ksl 152

labelling: and animal welfare 144, 147, 149, 150, 151, 152, 156, 157, 160; low calorie 18
land access: and community food initiatives 175–7, 181–2
land grabbing 18
land rights 19
landshare: farmer entrepreneurialism and rural community 179–81
Lang, T. 189, 195
Larner, W. 208
Lavik, R. 147
Lever, J. 160
Lewis, L.B. 196
LFB (London Food Board) 190, 193, 198
liberalization, food and market 68–73
Lidl 104
life cycle assessment (LCA) 96, 108, 114–15
local development 219
local food 47, 56, 115, 166, 193–4; supply chain 110; *see also* community food initiatives
local government 90
Lockwood, J.L. 125
London: Capital Growth initiative 175–6
London Food Board *see* LFB
London Food Link 175
Los Angeles 13, 188, 193, 195
Lösch, A. 31
low calorie labelling 18

McMichael, P. 11–12, 210
MAF (Ministry of Agriculture and Fisheries) (New Zealand) 130, 131–2; Quality Management 130–1
mainstreaming 25
Malmö 90, 188, 193, 198
malnutrition 9, 13
Malthus, Thomas 30
Malthusian catastrophe, prediction of 30
Marks & Spencer 107
Marsden, T. 6, 111, 169, 171, 207
Marx, Karl 4

mega-farm developments 19
metabolic rift 4, 10
methane 42
metrics 95
Metro 104
Mexico: overweight individuals 13
Miele, M. 111, 160
milk quotas 66
Mills, C. Wright 196
Millstone, E. 68
Ministry of Agriculture and Fisheries *see* MAF
Monsanto 217
Moore, J. 210, 211
Morgan, K. 11, 170, 199–200
Morris, C. 143, 160
Morrisons 107
multifunctionality: and agriculture 168–9; and farmer entrepreneurialism 167–71; and urban food strategies 192; in urban planning context 172
Murdoch, J. 111

National Animal Health Advisory Committee (NAHAC) (New Zealand) 131
National Farmers Union (NFU) 63, 136–7
National Lottery 99
National Minimum Wage Act 64
National Pest Management Strategy (NPMS) (New Zealand) 131
neo-productivism 9–10, 12; and urban planning 171–4
neoliberalism 8, 16, 18, 57, 87, 123, 127, 208, 211, 213; and animal disease 128–37; and biosecurity 127, 138–9; replacing of 214–19
Netherlands 19; animal welfare 148, 150, 155, 159
New Deal 6
New Food Equation 186
New York City 188, 192, 193, 195
New York City Council 190
New Zealand: Biosecurity Act 126; Bovine Tuberculosis 128–33, *129*, 134, 136, 137–8
Newquay 192, 195; Sustainability Strategy 195
Nitrates Directive (1991) 65
nitrogen fertilizers 39, *40*
nitrous oxide 42
no issue model 145; in Hungary 146
non-food diversification (NFD) strategy 170
Norgården 152
Norsvin: Action for Animal Welfare of (2001) 152
Norway: animal welfare 148, 152, 155, 156; farming 155–6
Nutrition Councils 189
nutrition transition 36–7, *36*; and water demand 38

obesity 9, 13, 18, 38, 86, 107, 189; cost to society of dealing with health implications of 86; number of people 86
OECD 45, 64, 86, 168–9
OIE (Office International des Epizooties) 126
oil 38–9, *39*
Oliver, Jamie 87, 90
Olympic Games Food Vision 190
Ontario 19
organic agriculture 47, *47*; environmental impacts 115
organic farmers and growers standard 149
organic food 56, 110, 111–14; chickens 111–12; interaction with conventional foods 114; moving to a system of 111–13; potatoes 113; and supermarkets 113; threats from agribusiness 114
overweight 13, 18, 38, 86
own label brands 105
Oxfam: *Growing to a Better Future* 56

packaging: environmental impact of 116
Party for Animals 148, 150
peak food 9
peak oil 209
Peck, J. 208
Pfizer case (2002) 74
Philadelphia 192, 193, 198
Plymouth 200
Poland 64
politics: and public sector 99–100
population 30, 33–8; agricultural 35, *35*; growth of 33–4, *33*; impact of growth in on food system 34; North and South differences 34; rural-to-urban transition 34
possums, control of 128, 131, 132–3
post-productivism 7–9, 10, 11, 12, 24, 169
potatoes, organic 113, 114
prices *see* food prices
private sector 117
productivism 6–7, 10, 11, 12, 24
property rights: reorientation of 18–20
public procurement 198

public sector 84–100; amount spent on food and catering services 84; barriers to sustainable food procurement 89–90; complexities of food procurement 89–93, 100; and EU public procurement regulations 85; evolution of catering 85–9; and Food for Life Partnership 87, 91, 92, 96; fragmented governance structure of 90; level of professional skills in 97; need for more innovative business models 97–8; outsourcing of services 97–8; percentage of GDP of purchase of goods and services by 84–5; and politics 99–100; school meals 85–9; and social enterprise models 99; values for money *see* values for money (public sector)
public–private regulation 207, 217

Quality Food from Hungary 153
quality of life 15

Ravenscroft, N. 172
Rayner, G. 194
ready meals 116
Redai, D. 146
reflexive governance 19; versus corporatist governance 16–18
refrigeration 115
Regional Accounts 216
Regional Animal Health Advice Committees (RAHACs) (New Zealand) 130
regulation: animal farming 144–5; public/private 207, 217
Renewables Directive (2009) 76
resources: depletion and exploitation of 3, 9, 10, 18
retailers 105–6; Code of Practice for 106; relationship with farmers 107; role of in food supply chains 104–5
Ring, P. 105
Roberts, Wayne 21
Romania 64
Rome, Treaty of 64
Royal Commission on Environmental Pollution 17
Ruhl, J.B. 77
Rural Development Plans (RDPs) 168
Rural Development Regulation (RDR) 168
rural–urban interface 166–83; and community food growing initiatives 174–81; multi-functionality and farmer entrepreneurialism 167–71; neo-productivism and urban planning 171–4

Sainsbury's 107, 206; environmental supply chain management 106
Sandwell Metropolitan Borough Council 91
Sandwell Primary Care Trust 91
Sanitary and Phytosanitary (SPS) Agreement 70
Sayer, Andrew 93–4
Sbjöblom, S. 22
Scaling up in Agriculture, Rural Development, and Nutrition 50
Scaling up Global Food Security and Sustainable Agriculture 50, 54–5
Scandinavia: welfare state model of animal welfare 147–8; *see also* Norway; Sweden
Schiff, R. 189, 196
Schneider, S.A. 77–8
school meals 85, 85–7, 90, 92, 97, 100, 200; barriers to reform 90; ecological model 87–8; whole school approach 87; and Whole Schools Meals Ltd (WSM) 98–9
Schot, Johan 109–10
science-based approach: European food governance 73–6
Scientific Committee on Food 70
Scotland: Hungry for Success initiative 87, 91
Self, P. 63
'shopping trolley' study 114–15
short-termism 21–2, 57–8
Single European Market 8, 62
Single Payment Schemes (SPS) 168
Sjöblom, S. 21
Skogstad, Grace 63
Slow Food 111
small-scale farming, importance of 52
Smith, Adam 94
social enterprise models 99
social return on investment (SROI) 95–6
social sciences: ethical turn in 93–4
Sodexho 91, 97
Soil Association 56, 87; Food for Life initiative 200; organic standard 149
soil science 4
Sonnino, R. 171, 186
Southerton, D. 109
Soviet Union 6
Spaargaren, G. 11
spatial planning 197–8
spatial policies: temporal mechanism in 21–2
stakeholders 53; engaging with food policy councils 191, 199; involvement in urban food strategies 199–200

Standing Committee on Agricultural Research 23
state: and agriculture 6; and food system 117
statistics 31; fallibility of 31
Stolle, D. 145
Storey, J. 105
Storing, H. 63
Stroud Farmers' Market 178–9
Stroudco community food hub 177–9
supermarket model 145, 148–53
supermarkets 104–5; dominance of food system 104; and environmental supply chain management 106–7; and organic food 113, 114; own label brands 105; and reduction of waste 107
Sustain 87, 191; *Good Planning for Good Food* 200
sustainability 3–7, 9, 14; evidence of uptake of 47–8; and food security 15–16, 20, 23–4; key parameters for defining systems of 14, 15; and short-termism 21–2
sustainability crisis 9–10, 11
Sustainable Development Commission 17
Sustainable Food City Network 200, 201
sustainable food paradigm: towards a new conceptual architecture for a 14–22
sustainable intensification approach 170
Sustainable Procurement Task Force 97
sustainable rural development 169
Swanwick, C. 173
Sweden: animal welfare 147–8, 151–2, 155, 156; cattle sector 155
Swinback, A. 67

Taylor, B. 172
terroir model 145, 147
Tesco 104, 107
third nature thinking 10
Toronto 19, 188, 189, 195; food policy council 189, 196; food strategy 13, 198; health-focused food system 192–3
Toronto Public Health Department 190
Towards the Future We Want 50, 52
Town and Country Planning Act (1947) 5, 172
trade barriers, reducing 51
transgenic crops 45, 52
Transition Theory 109–10
transport, environmental impacts of 115
Tregear, A. 111
Tronto, Joan 94

UFSs *see* urban food strategies
undernourishment: decrease in global levels 37–8; statistics on 31
United Kingdom (UK) 19; animal welfare 148–9, 155; Bovine Tuberculosis 128–9, *129*, 133–7, 138; relationship between state and farmer 63–4; school meals provision 86–7
United States 6; biotech crop production 45, *45*; and obesity 38, 86; R&D funding for agriculture 47; school meals programme 88–9
Universities 16
urban agriculture 19
urban food governance: tapping the potential of 197–8; transformative potential of 194–7
Urban Food Strategies (UFSs) 16, 188, 192–4, 195, 197, 199; and multifunctionality 192; research framework for sustainable 201–2
urban planning: and inclusion of greenspace 173; and neo-productivism 171–4
urban strategies (for food security and sustainability) 186–202; adoption of comprehensive policy approach 199; criteria for sustainable 199–200; emergence of cities as food policy actors 187–9; food security and sustainability 186–202; harnessing cultural change 200; infrastructural development 197; and involvement of stakeholders 199–200; local food system 193–4; monitoring and evaluation system 200; need for clear vision 199; public procurement 198; spatial planning 197–8
urbanisation 4
Utrecht (Netherlands) 190

value-adding (VA) strategy 170
values for money (public sector) 93–100; enabling 97–100; measuring 95–6; recognising 95; within reason 93–4
Van de Ven, A. 105
van Zwanenberg, P. 68
Veerman, J.L. 77
Vision 2050 50, 53–4
Vogel, D. 71

Wales 19, 78; Ecoli crisis (2005) 100; procurement skills deficit 97; school meals 88
WalMart 104

waste food 21, 53, 117, 194, 199; supermarkets and reduction of 107
water: environmental controls to protect 65; nutrition transition and demand for 38; used for agriculture and food production 38
Welfare Quality Protocols 158
welfare state model 145, 147–8
well-being 13, 15
WHO (World Health Organization) 86
whole life costs 96
whole school approach: and school meals 87
Whole Schools Meals Ltd (WSM) 98–9
Wickson, F. 74
Wildlife Trust 135
Williamson, M. 125
World Bank 42, 209; Food Price Index 32
World Business Council for Sustainable Development (WBCSD) 53, 56
world ecology 212, 213; food as part of 206–9
World Society for the Protection of Animals (WSPA) 146
WTO (World Trade Organization) beef hormones dispute 72; Sanitary and Phytosanitary (SPS) agreement 126; Uruguay Round 32–3, 65
Wynne, B. 74

yields 211–12; closing productivity gap 52
Yong Kim, Jim 209